Environmental Communication Pedagogy and Practice

Given the urgency of environmental problems, how we communicate about our ecological relations is crucial. *Environmental Communication Pedagogy and Practice* is concerned with ways to help learners effectively navigate and consciously contribute to the communication shaping our environmental present and future.

The book brings together international educators working from a variety of perspectives to engage both theory and application. Contributors address how pedagogy can stimulate ecological wakefulness, support diverse and praxis-based ways of learning, and nurture environmental change agents. Additionally, the volume responds to a practical need to increase teaching effectiveness of environmental communication across disciplines by offering a repertoire of useful learning activities and assignments. Altogether, it provides an impetus for reflection upon and enhancement of our own practice as environmental educators, practitioners, and students.

Environmental Communication Pedagogy and Practice is an essential resource for those working in environmental communication, environmental and sustainability studies, environmental journalism, environmental planning and management, environmental sciences, media studies, and cultural studies, as well as communication subfields such as rhetoric, conflict and mediation, and intercultural. The volume is also a valuable resource for environmental communication professionals working with communities and governmental and non-governmental environmental organizations.

Tema Milstein is Presidential Teaching Fellow at the University of New Mexico, USA. She is an Associate Professor in the Department of Communication and Journalism, and Affiliate Faculty in the Geography and Environmental Studies Department, as well as the Sustainability Studies Program.

Mairi Pileggi is an Associate Professor in the Department of Communication and Media Studies, and Director of the Gender Studies Program at Dominican University of California, USA.

Eric Morgan is an Associate Professor in the Department of Communication Studies at New Mexico State University, USA.

Routledge Studies in Environmental Communication and Media

Culture, Development and Petroleum
An Ethnography of the High North
Edited by Jan-Oddvar Sørnes, Larry Browning and Jan Terje Henriksen

Discourses of Global Climate Change
Apocalyptic framing and political antagonisms
Jonas Anshelm and Martin Hultman

The Troubled Rhetoric and Communication of Climate Change
The argumentative situation
Philip Eubanks

Environmental Communication and Travel Journalism
Consumerism, Conflict and Concern
Lynette McGaurr

Environmental Ethics and Film
Pat Brereton

Environmental Crises in Central Asia
From steppes to seas, from deserts to glaciers
Edited by Eric Freedman and Mark Neuzil

Environmental Advertising in China and the USA
The desire to go green
Xinghua Li

Public Perception of Climate Change
Policy and communication
Bjoern Hagen

Environmental Communication and Community
Constructive and destructive dynamics of social transformation
Tarla Rai Peterson, Hanna Ljunggren Bergeå, Andrea M. Feldpausch-Parker and Kaisa Raitio

Environmental Communication Pedagogy and Practice
Edited by Tema Milstein, Mairi Pileggi and Eric Morgan

Environmental Communication Pedagogy and Practice

Edited by
Tema Milstein,
Mairi Pileggi, and
Eric Morgan

Routledge
Taylor & Francis Group
LONDON AND NEW YORK

earthscan
from Routledge

First published 2017 by Routledge

2 Park Square, Milton Park, Abingdon, Oxfordshire OX14 4RN
711 Third Avenue, New York, NY 10017

Routledge is an imprint of the Taylor & Francis Group, an informa business

First issued in paperback 2018

British Library Cataloguing in Publication Data
A catalogue record for this book is available from the British Library

Library of Congress Cataloging in Publication Data
Names: Milstein, Tema, editor. | Pileggi, Mairi, editor. | Morgan, Eric (Eric L.), editor.
Title: Environmental communication pedagogy and practice / edited by Tema Milstein, Mairi Pileggi and Eric Morgan.
Description: New York : Routledge, 2017.
Identifiers: LCCN 2016037852 | ISBN 9781138673090 (hb) | ISBN 9781315562148 (ebook)
Subjects: LCSH: Communication in the environmental sciences. | Environmental sciences–Study and teaching. | Communication in education.
Classification: LCC GE25 .E578 2017 | DDC 363.70071–dc23
LC record available at https://lccn.loc.gov/2016037852

ISBN: 978-1-138-67309-0 (hbk)
ISBN: 978-1-138-39350-9 (pbk)

Typeset in Goudy
by Wearset Ltd, Boldon, Tyne and Wear

Contents

Figures

Tables

Contributors

Maryam Alhinai, a PhD candidate at the University of New Mexico, focuses on ecocultural communication. Maryam holds an MA from Monterey Institute of International Studies, California, in teaching English to speakers of other languages. As a lecturer at Sultan Qaboos University, Oman, she teaches English communication skills.

Donal Carbaugh is Professor of Communication at the University of Massachusetts Amherst and Distinguished Scholar of the National Communication Association, its highest honor. His interests are to develop an understanding of communication that brings together nature and spirit in cross-cultural perspective as well as the development of cultural discourse analysis.

José Castro is a PhD candidate at the University of New Mexico. In his classes and research, he bridges ecocultural, environmental, and intercultural communication to explore cultural frames constructing the environment. He co-creates the classroom as a space for activism, social change, and environmental justice for both humans and nonhumans.

Katherine M. Cruger, Assistant Professor of Communication at Chatham University, directs M.A. programs in Professional Writing and Communication. Her research agenda includes rhetorical constructions of fracking by industry rhetors, sustainability pedagogies, and improving conditions of new parenthood. She lives in Pittsburgh, PA, with her partner, son, and many plant and animal companions.

Jake Dionne (M.A. Syracuse University) is a PhD candidate in the Department of Communication at the University of Colorado, Boulder. His research and pedagogy comments on the need to consider and show consideration toward nonhumans, especially those other-than-human beings enduring physical erasure due to extinction.

Caroline Gottschalk Druschke is an Associate Professor in the Departments of Writing & Rhetoric and Natural Resources Science at the University of Rhode Island, where she directs the Society, Ecology & Communication Lab. Her training in rhetoric and ecology informs work focused on human dimensions of natural resources management.

Linda Flower is a Professor of Rhetoric at Carnegie Mellon. Her research in writing as a social-cognitive process laid a foundation for teaching writing as the art of rhetorical problem-solving. Working with college students, professionals and marginalized communities has extended this concern with rhetorical agency and social action to intercultural inquiry within deliberative local publics.

Carrie P. Freeman, Associate Professor of Communication at Georgia State University, publishes and teaches on media ethics, activist communication, environmental communication, and critical animal studies. She wrote the 2014 book *Framing Farming: Communication Strategies for Animal Rights* (www.framingfarming.com) and co-authors the www.animalsandmedia.org style guide. A grassroots activist for several decades, Freeman currently co-hosts animal and environmental protection shows on Radio Free Georgia.

Stephen Griego is a PhD student in the Communication and Journalism program at the University of New Mexico. His work focuses on the relationships among geography, food as communication, and cultural identity. As a student of Dr. Tema Milstein, Stephen has added a more ecocentric standpoint to his research.

Gabriele Hadl is Associate Professor at Kwansei Gakuin University and Secretary of the International Environmental Communication Association (theIECA.org). She has recently published a book on EcoMedia Literacy (in Japanese) and believes in all generations' rights to a starry sky and duties to protect trees ("window-rights and tree-duties" F. Hundertwasser).

Lars Hallgren is Assistant Professor and Senior Lecturer at the Division of Environmental Communication, Department of Urban and Rural development, Swedish University of Agricultural Studies, Uppsala, Sweden. Hallgren has been involved in developing and teaching the two-year master of science program "Environmental Communication and Management" since 2006.

Joy Hamilton is a PhD student at the University of Minnesota's Department of Communication Studies. She recently served as a research assistant for the Provost's Experiments in Learning Innovation initiative, which "flipped" the environmental communication course from inside the classroom to a digitally networked field course using public land sites.

Karey Harrison is a Senior Lecturer in Communication at the University of Southern Queensland where she recently received Awards for Teaching Excellence. Her teaching and research focus on the metaphorical structuring of institutional discourses of science and technology. She is also a presenter in Al Gore's Climate Project.

Jeff Hoffmann is a member of the 2014 PhD cohort in Intercultural Communication at UNM. His research interests include intercultural communication, ecocultural and environmental communication, sustainability, political ecology, conflict and food justice, and language acquisition.

Sigrid Kannengießer is a Research Associate at the Center for Media, Communication and Information Research at the University of Bremen, Germany. Her research and teaching focuses on environmental communication, media sociology, media and political movements, media and materiality, transcultural communication, and gender media studies. For more information visit: www.zemki.uni-bremen.de.

Laura Lindenfeld is Director of the Alan Alda Center for Communicating Science and a Professor in the School of Journalism at Stony Brook University. Her environmental communication and sustainability science research and teaching focus on how individuals and institutions can create stronger linkages between the production of knowledge and action.

Antonio López, PhD, has a research focus on bridging sustainability with media literacy. He has written numerous academic articles, essays, and books. His most recent book is *Greening Media Education: Bridging Media Literacy with Green Cultural Citizenship* (Peter Lang). He currently teaches at John Cabot University in Rome, Italy.

Bridie McGreavy is an Assistant Professor in the Department of Communication and Journalism at the University of Maine. She studies how communication shapes relationships between humans and marine and freshwater ecosystems. Her research and teaching in environmental communication and sustainability science seeks to build healthy and resilient communities.

Tema Milstein, an Associate Professor and Presidential Teaching Fellow at University of New Mexico, has taught at universities in the United States, Europe, and Asia. Her research examines ways that communication as a cultural force informs human relations with/in/as nature.

Gabi Mocatta joined the University of Tasmania after a fifteen-year career in print journalism during which she often covered the environment. She currently teaches online and face to face, and researches environmental activism, corporate social responsibility, and "green-washing." She is author of the new online Open School of Journalism's course in Environmental Journalism.

Eric Morgan is an Associate Professor of Communication at New Mexico State University. His research interests include cultural discourses of environment and the ethnography of communication. When not working, he can be found outside climbing, hiking, and/or playing music with friends and family.

Melissa M. Parks earned her BA in creative writing at the University of California, Riverside (2008). She is an international teacher and served as a Peace Corps Volunteer in Malawi, Africa (2011–13). She completed her MA in ecocultural communication at the University of New Mexico and is currently a PhD student at the University of Utah.

Mark Pedelty is a Professor of Communication Studies at the University of Minnesota and Resident Fellow at the Institute on the Environment. His recent books include *Ecomusicology: Rock, Folk, and the Environment* (Temple) and *A Song to Save the Salish Sea: Musical Performance as Environmental Activism* (Indiana University Press).

Aaron T. Phillips is an Assistant Professor in the department of Management at the University of Utah. His research focuses on writing as a social practice, the perception/experience of risk, and the development of (eco)literacy.

Mairi Pileggi is an Associate Professor of Communication and Media Studies at Dominican University of California. A storyteller, she approaches environmental communication through a feminist lens. An avid gardener and cook, she enjoys walking the beach with her family.

Emily Plec is a Professor of Communication Studies at Western Oregon University and editor of the book *Perspectives on Human–Animal Communication: Internatural Communication*. Her teaching and scholarship addresses the intersections of communication and social justice with an emphasis on rhetorical, environmental, cultural, and media studies.

Jessica M. Prody is an Assistant Professor of Performance and Communication Arts at St. Lawrence University where she teaches Rhetoric and Environmental Communication courses and researches environmental citizenship, climate change, and the local food movement.

Maggie Siebert, a PhD candidate at University of New Mexico, incorporates health and environmental communication lenses in her research on food, from genetically modified organisms to local food systems. Maggie holds a BA in Cultural Anthropology, an MS in Community Health, and teaches courses in communication and sustainability studies.

Stacey K. Sowards is a Professor and Chair of the Department of Communication at the University of Texas at El Paso. She is also the coordinator of the Rare program at UTEP.

Leah Sprain is an Assistant Professor in the Department of Communication at the University of Colorado Boulder. Her research focuses on democratic engagement, studying how communication practices shape democratic action. Her research and teaching draw on language and social interaction perspectives to explore environmental communication, deliberation, and social movement activism.

Garrett Stack is a PhD candidate at Carnegie Mellon University. He has written a chapter in the forthcoming *Envisioning Futures for Environmental and Sustainability Education*, part of Wageningen Academic Publishers' United Nations Decade of Education for Sustainable Development book series, and his dissertation focuses on the Sierra Club and its role in establishing a set of rhetorical methods for arguing in favor of preservation of American wilderness.

Mary Stroud is a lecturer in the Program in Writing and Rhetoric at Stanford University with research interests in environmental rhetoric, new media studies, and humor studies. She holds a PhD in rhetoric and composition from the University of Arizona and has published works in the fields of environmental pedagogy, composition theory, and organizational leadership in secondary and higher education.

Geo Takach is a writer, filmmaker, speaker, and instructor. His adventures span hundreds of publications among speeches, print, theatre, film, radio, television, corporatedom and Boolean ether. He is an Associate Professor and Head of the BA in Professional Communication online program in the School of Communication and Culture at Royal Roads University in Victoria, Canada.

Carlos A. Tarin is Visiting Assistant Professor in the Department of Communication at the University of Texas at El Paso. His research focuses on the intersection of environmental and organizational communication, particularly in international contexts.

Mariko Thomas is a PhD student at University of New Mexico studying eco-culture, environmental communication, and intercultural communication. Her specific areas of study include social justice and social change, identity, place, space, and movement through an ecological lens, and issues of voice and power in relation to environmental access.

Jessica L. Thompson is an Associate Professor in the Communication and Performance Studies Department at Northern Michigan University. Her research is focused on communicating and managing environmental and organizational communication about complex ecological and social issues. In her classes, she focuses on communicating global climate change and designing public relations campaigns for natural resource conservation.

Eli Typhina is a Communication, Rhetoric, and Digital Media doctoral candidate at North Carolina State University. Since 2002, Typhina has taught and designed hands-on environmental curricula for local, national, and international outdoor education centers, environmental agencies, and universities. In 2015, Typhina received the North Carolina State University teaching excellence award.

Sarah D. Upton is Assistant Professor in the Department of Communication at the University of Texas at El Paso. Her research focuses on engaged communication, exploring themes such as invitational social change for community development, communication for environmental conservation, relationship building and transformation in qualitative research, and service-learning pedagogy.

Maria Clara Valencia is a Journalist and Professional in Literary Studies from Colombia dedicated to the environment. She leads classes on environmental journalism at Universidad Tecnológica de Bolívar in Cartagena de Indias (Colombia) and also has a research group on the topic of environmental journalism.

Grady Walker received his PhD from the Centre for Communication and Social Change at The University of Queensland, Australia. He also holds a master's degree from New York University in environmental conservation education. Previously he worked internationally as a documentary filmmaker.

Kenneth C. C. Yang is a Professor in the Department of Communication at the University of Texas at El Paso. His research focuses on new media, consumer behavior, and international advertising. Some of his many works have been published in *International Journal of Consumer Marketing*, *Journal of Intercultural Communication Studies*, and *Telematics and Informatics*.

Foreword

Donal Carbaugh

Like a diverse bio-region, this book offers a rich range of eco-discourses for reflection and cultivation. As the editors discuss, there are a variety of fertile grounds for reorienting ourselves, novel areas for pedagogical practice, specific tactics for learning anew about relations with environments, and important ethical considerations to consider. As such, the book offers rich grounds for reflection while helping to cultivate our studies of environmental communication. I add a few thoughts in this foreword as a way of contributing to these important efforts.

One central feature in the study of environmental communication can be understood by asking, what is its place? I use the word, place, deliberately here and want to exploit its various meanings. There is place as an actual, physical environment. What, exactly, or where, are we thinking about? What place or places do we have in mind? This matters as we fix our thoughts, here and there, as we think through places about our communication, and think through our communication about places.

Place is also a symbolic space. We weave meanings with words, thereby creating a verbal rendering of place, always an exercise in partial and selective attention. One might find it useful to mention the grove of hemlock trees and the ways hemlock alternates with birch and an occasional beech. An earlier period would have supported chestnuts, but this has given way today to a few mere saplings, which soon give way to blight. One might also mention avian life, the vibrant role of the nuthatches and their tear-tear-tear sounds. Echoes of the chickadee animate the soundscape with their dee-dee-dee. The spring dusk and dawn are good times to be treated to the winter wren's long warbling and cheerful song. Similarly the wood thrush brings sounds of joy to the woods.

We have here, then, at least two types of places brought into view. One is the emphatic physical environment in which we inevitably dwell. Our worlds are material places in which we live; we cannot escape being emplaced in this sense. As Henry David Thoreau said long ago, we are "part and parcel of nature." And it matters deeply where we are. Thoreau's "sauntering" took him to the Concord woods and Walden Pond. Edward Abbey's desert, John Muir's wonderful valley in the Sierra Nevadas, a Tibetan's steppe, a Lapp's northern fell, and your places, too. It matters deeply where, in what places, we are.

It matters deeply also where we imagine or think we are, or express ourselves as being. One form of this imagination is spun with our words. Above, we "say" we are part of a mixed coniferous (hemlock) and deciduous (birch, beech) forest. Or, alternately, we say we are dwelling with our feathered friends. These words provide a symbolic rendering, which we lay over our place—including its sounds—as we dwell within it.

The point of this distinction between the material and the symbolic is the necessity of developing some critical capacity relative to both of these places: the literal physical space, and the linguistic discourses about them. Too often, we harness or enslave a physical place with one discourse, via selective and partial vision, and then treat the discourse as standing as a whole for what that place is, was, or could be. In the process, we can find that we are treating the physical environment with primary attention to our exclusively worded world. We can enslave our environment within a prison-house of a discourse. The symbolic can overtake "the actual" as we treat our words as being actually what is indeed there.

There, that place, no doubt could offer us new insights, possibilities beyond our worded symbolic rendering. Perhaps the chemist links toxic compounds to a place, diagnoses the ill effects, and newly explains environmental harm in the process. This one discourse can set the stage, potentially, for useful interventions in places, the symbolic creating different ways of intervening. But this also might risk, in some places, a reduction of a place to chemical compounds. Balance between the material and symbolic is always needed. Just as birds need trees, compounds need organisms, all of which need bioregions and ecosystems. We might think of each—physical particle, chemical, organism, bioregion—as a level of a system, or metaphorically as a rung on a ladder. As we move from particles to chemicals to compounds to organisms to bioregions and back again. There is a ladder of understanding here, or similarly a nested conceptualization of matters usefully exploited here. Each time we move up or down the ladder of discourses, from within or outside the nest, we are offered a new perspective.

This all can be mindfully reflected upon further as spiritual matters, infused with ancestral wisdom, carrying lessons loudly to some but for others falling on deaf ears. The symbolic possibilities can be, seemingly, endless. This all depends on what place(s), then, we are considering and with which words (pictures, sounds) we are considering—for the time being. There are nearly endless possibilities in how our places can be linguistically coded.

Take, for example, a historical practice in Finland known in Finnish as *metsänpeitto*. This term characterizes a state of being when one treats a wooded area disrespectfully. Perhaps you were too noisy or mistreated a tree or the water, in which case the little spirits of the forest might overtake you and confuse you to the point of overwhelming fascination, leading to you surrendering deeply to your surroundings. In this state, supernatural insights may be revealed to you. It is important to emphasize that this forest experience, despite the temporary disorientation, is not all bad. *Metsänpeitto* also introduces the possibility of being in a joyful way at one with the forest.

The point is that each language carries multiple possibilities for rendering our world to us: each is distinctive in what it offers about places; each also is housed in places of its own making—in the case above, Finland in particular.

Because of this, the experiencing of our emphatically real world and our symbolic renderings of it must be given some parity of expression. This important point requires hearing not only our own words about our worlds and others' words about our worlds, but also hearing in the other direction as our worlds express themselves to us in their own ways. Communication goes—or should be understood to go—both ways.

When we fail to reflect on this process, refuse to cultivate variety, don't hear both our words and our worlds, our places can suffer. When we cannot critically distance ourselves from our preferred words about our places, thereby grounding ourselves in the places of our words, we can set ourselves up for failure.

As an invested participant in my own places, I have learned difficult lessons in listening to various systems of expression—I call them cultural discourses—each with their own vision (voices, verbs, sounds) of a place, often with competing visions, with the competition getting intense. One example is the place where gas, oil, and mining leases were granted by the United States government into Native American lands, such as the Badger-Two Medicine area of Blackfeet Territory in northern Montana. Extracting parts of our places because it is possible does not stand easily within a spiritual landscape, long a site of sacred ceremonies. Just as an ecosystem benefits from and requires diversity, so too does our understanding of some places benefit from diverse cultural as well as physical geographies. Multiple discourses—of trees, birds, chemicals, spirituality—lend distinct insights into our places, all requiring consideration as the sites of which we are inevitably only a part.

Yes, place matters, both physically and symbolically. And, while the meaning of each can be separated, and must be separated for purposes of understanding and critical reflection, as indeed I have done above, these are matters intimately and organically tied one to another. We must search our places to see–hear–sense what they tell us. This is very likely how the chemist found what she called "the toxins." In turn, we must search our discourses for what they say; in the process, we find what they miss. There is always something different and better to say. This sketch I take to be the lifeblood of environmental communication.

The editors ask, in the first unit of this volume, what orientation does one take to learning about environmental communication, and where might that place one as a learner or teacher? There are many possibilities. If we use the natural landscape metaphor, we could sit on a lakeshore, watching the intersection of water and land, water creatures as well as those scurrying the shoreline. We might sit deep in the forest, attentive to plants, animals, visual and aural spheres activated but perhaps in a dim light. We could perch upon a rock outcrop on a ridgeline, possibly seeing far distances. We could meditate upon the spiritual dimension as Thoreau, Emerson, Tolstoy, St. Augustine, Muir, and others have done before us. Places offer different orientations from which we can study, reflect, and learn critically. And likewise in the study of an environment or place we weave our meanings with words. Our words lead us through our places. And, in turn, the places can lead us to new, better words.

It is fruitful to dig deep into these matters, as do the contributors to this volume. An additional way of doing this is to ask two fundamental questions:

- Where is (environmental) knowledge?
- How is that knowledge being communicated?

Most students of places, or environments, have rich experiences in them. They have walked trails, rafted rivers, climbed up or down cliffs to see raptors, or the like. These experiences in places animate their reading about forestry, hydrology, or ornithology. In my classes when I ask the question, where is environmental knowledge, most say that knowledge resides in books, articles, or studies, each a product of a way of knowing typical of a Westernized linguistic perspective. This has no doubt been resoundingly powerful in countless ways. At the same time, I ask students about our campus, or the woods that they are in and that surround us; they don't really know the place. What I am trying to say is the place does not "speak" to them as they dwell within it. Why is this? They are apparently unable (or perhaps, if able, unwilling) to discourse about the trees, birds, and so on around them. Occasionally, there is a hiker in class who can utter a name of a local trail.

What do these students know? What don't they know?

I have found myself in much sympathy with Louv's claim that many of us suffer from a "nature deficit disorder." My way of addressing this is to assign Thoreau's essay on "sauntering," discuss it in great detail, then send students to the woods for a good long saunter. I suggest their reading for the week is not in the book or articles, but in the woods. Go to that text, study it in as much detail as you can, then, when you come back, let's discuss what each of you has learned. This is a favored exercise of mine—and the authors in the final part of this book provide many of their own. The discussion includes themes of obstacles (can I saunter with headphones?; I kept thinking about my homework), which we discuss and address (no, no headphones; why didn't this seem like homework?). It also includes testimony of links to places in new ways. Now, I think, we are getting somewhere—physically and symbolically. The students now have knowledge they did not have before. Most of them like it. Sometimes there is even expression of awe and wonder. In the end, some students say they will never walk across campus in the same way again.

Then, we read John Muir's essay about the water ouzel. We ask what knowledge he had to have to produce the prose in that essay. This leads to them imagining being Muir, on a creek bank, watching in the way he must have watched to see the world before him—as it expressed itself to him. Students struggle to hear the world in that way, but the exercise opens new possibilities to them. In the process, they may link Muir's experiential and linguistic abilities to Thoreau's type of sauntering. In each case, "being there" in the wild is tied intimately to knowing more deeply what one is writing, and reading. Knowledge, in other words, is tied to the world's expressions of itself to us.

So, knowledge, or a special type of it, wisdom, as Keith Basso titled it, can be understood as it "sits in places" (Basso, 1996). We owe it to ourselves, our places, and our students, to critically reflect upon such knowledge about environments and what we are saying about them; but, moreover, we owe it to our places to open ourselves to what they are saying to us. This means, to the extent we can, being there (I find wheelchair accessible places when needed), and learning about the place(s) as in a practice of Thoreauvian sauntering. It means also sorting through possible discourses about places and studying those, but always as attached to those places (the world being a special case itself). We should not let the symbolic part of our discourses run well above the places we seek to know. We can do a disservice to all when we do.

The editors of and contributors to this wonderful volume have brought to you a rich and varied range of discourses, ways to orient to environments, approaches to teaching about environmental relations, as well as the ethical dimensions at play when we communicate about our places. Nicely specific activities of learning in and out of classrooms, as discussed herein, help us develop such knowledge. Together, we seek to understand better what is involved in environmental communication. This volume helps move us along those various paths. Enjoy the journey.

Reference

Basso, Keith H. (1996) *Wisdom Sits in Places: Landscape and Language Among the Western Apache*, University of New Mexico Press, Albuquerque.

Introducing environmental communication pedagogy and practice

Tema Milstein, Mairi Pileggi, and Eric Morgan

Many books regarding the current ecological health of Earth begin with a list of the anthropogenic, utterly systemic devastations spilling from and upon us. Pedagogues and practitioners of environmental communication take seriously engaging ourselves and others with these phenomena and understanding our implications in them. Current ecological and cultural exigencies can be experienced as a rising swell of aching, questioning, caring, and pushback, a welling up from source, and a gift of movement surging forth. In the ensuing flow, pulsing through borders, species, bodies, oceans, and atmosphere, the urgency for conscious and skillful communicators is clear. As such, this book is about pedagogy concerned with ways to engage learners in ecological wakefulness so we may avoid sinking speechlessly beneath the surface and instead work to effectively navigate and consciously contribute to generative discourse and praxis.

This book marks a first attempt to put into conversation our work as educators of environmental communication. As teachers, we facilitate contending with how to make sense of our roles in ecological phenomena, how to respond individually and as communities and institutions, and how to enable ourselves and nurture others to be agents of restorative change. While for some time we have been conducting this work on our own, we thankfully have had a robust literature offering guiding content, including textbooks such as Cox and Pezzullo (2016) and the previous editions of the text Cox authored alone, Corbett (2006), Hendry (2010), Stibbe (2015), and Hansen (2010), as well as a continually growing abundance of useful articles, chapters, and books.

Our solo journeys into teaching often have been paired with designing the first environmental communication courses offered by our institutions—often the first such course we have taught or, for that matter, taken. We have frequently produced creative and innovative pedagogy, refreshingly unfettered by conventions or established disciplinary expectations. However, this freedom at times comes with many of us thinking we must start from scratch, feeling unsupported or alone, and taking avoidable missteps. This book provides a raft of support from others who have gone before or worked parallel to us, one that can assist and propel us in our voyage.

Our book was inspired by those of us who, looking for just such support, organized a gathering of environmental communication educators and practitioners to discuss pedagogy and practice at the annual National Communication Association meeting in November 2014 in Chicago, Illinois, USA. The response was overwhelmingly positive, and reports from those there—and those who heard about the gathering—indicated a deep desire to continue and to engage a larger global community in the conversation. This resultant book's contributors teach in different parts of the world and the chapters and activities are diverse, accessing a range of teaching philosophies, though tending toward constructivist, performative, and radical approaches. Hailing from large public universities to small private colleges, the contributors work with adult learners, including undergraduate and graduate students, community members, and professionals. Although many teach largely in classroom settings, some of their teaching also happens on city streets, in forests, and in boardrooms. Together, these pedagogical orientations and conditions provide impetus for reflection on, and enrichment of, our work as educators and practitioners of environmental communication.

What do we mean by environmental communication?

Environmental Communication as a field of inquiry intersects with a variety of academic disciplines within the social and natural sciences, the humanities, and environmental studies. The field's research hinges on two core assumptions:

- the ways we communicate powerfully shape our ecological perceptions; and
- these perceptions inform how we act with/in the more-than-human world (Milstein 2009).

As such, communication is understood as a powerful social process that not only reflects but also does work in "constructing, producing, and naturalizing" particular ecological relations and realities (ibid., 345). Our book embraces these assumptions, as well as another widely held assumption in the field: Human representations of the biosphere are *interested*. Namely, these representations are informed by social, economic, and political interests embedded in contexts that allow for certain ecological perceptions and that obscure others.

Further, much of this book aligns with environmental communication scholarship that challenges conventional academic orientations, which tend to reproduce dominant Western premises of "nature" as a separate, passive backdrop or a mute store of resources. In doing so, we attempt to eschew human exceptionalism in our teaching, as well as our research, and trouble the dominant Western culture–nature binary associated with anthropocentrism. We also seek to understand and articulate ways the more-than-human world mediates communication and culture. Congruent with all these assumptions and moves is an ethical imperative championed by some leaders in the field who view environmental communication as a crisis discipline. In this vein, environmental communication scholar Robert Cox (2007), former three-time president of the Sierra Club, argues inquiry addressing phenomena such as toxic pollution, mass extinction, and climate disruption has an ethical duty to not only illustrate and explain but also to help change those aspects of society causing ecological collapse. These assumptions and imperatives are the ground of environmental communication pedagogy and practice.

What is environmental communication pedagogy?

This book contends with the basic question of how environmental communication pedagogy embodies and imparts the contexts, assumptions, and principles laid out above. Environmental communication pedagogy provides abundant opportunities to rethink the foundations of, ways to, and reasons for teaching. In conventional higher educational contexts, knowledge construction has been confined most often to a classroom constrained by power and authority granted the instructor and the institution. The hallmark of such teaching and learning generally engages a rational approach to the world that privileges abstraction. Certainly, many educators have taken exception to this mode of pedagogy, most fundamentally in overhauls of the core philosophy of teaching seen in problem-posing and consciousness-raising approaches of radical and critical pedagogy and the ecoculturally transformative approaches of ecopedagogy.

Environmental communication invites teachers and learners alike into a different epistemological space—one where knowledge and learning is not utterly dominated by abstract theoretical notions but where it is material, interconnecting, affecting, and urgent. Just as environmental communication as a field of inquiry challenges not only conventional academic but also cultural norms, each contributor to this volume in some way seeks to challenge traditional or time-worn ways of engaging in pedagogy.

Most of the book's contributors started environmental communication courses as special topics before they had a place in course catalogs or as orientations within existing general courses that took on an environmental lens to communication concepts/theories/frameworks. Many of these then became permanent environmental communication courses at institutions, making way for recognition, legitimacy, and support as part of the established curriculum and breaking trail for others who would follow in their footsteps. Students have been quick to respond, with course waiting lists, a general buzz, and deeply felt, and fed, passion.

Contents of this book

Whereas, from a theoretical standpoint, this book engages discussions of the qualities, principles, and goals of pedagogy, this book is equally about practice. This places the focus, in part, on how we can engage learners in processes of transformation so together we might swiftly and consciously aid in transforming dominant modes of engagement with our world. Many of the chapters provide core praxis frames, which will be useful for teachers who may have the theory and concepts clearly taught, but need ways to help learners put those ideas into the practice of being effective agents of change. Equally, practitioners of environmental communication may find the questions, concepts, frameworks, and learning tools shared in these pages provide a useful space for reflection on, and enhancement of, their ongoing work.

This book is partitioned into four interrelated parts. Authors in **Part I, (Re)conceptualizing the environmental communication classroom**, focus on ways educators are creatively adapting pedagogy to engage learners in environmental communication and how teachers can (re)orient and (re)locate the classroom. In **Negotiating knowledge and advocating change: teaching for transformative practice**, Stack and Flower explore ways students learn as *knowledge-negotiators* to encounter and to analyze ecological issues as they strengthen their communicative abilities and as *transformative public advocates* to produce genuinely useful contributions to current environmental issues. This teaching approach moves from a Habermasian "salon" that privileges communicative skill building and open negotiation of a wide variety of environmental topics, to a more complex model where strategic communicative skills and emerging opinions become a foundation upon which students base their own arguments, deepen their knowledge, and ultimately feel competent to enter the public environmental domain as advocates. Shifting the focus from students to teachers, Prody argues that environmental communication educators are indeed practitioners, and so must identify their purpose, account for their audience, and adapt to particular contextual constraints. In **Pedagogy as environmental communication: the rhetorical situations of the classroom**, she reflects on her experiences teaching three very different environmental communication courses, highlighting ways to effectively evaluate and adapt one's practice to successfully support students in adopting ecological citizenship. In **Environmental communication pedagogy: a survey of the field**, Hamilton and Pedelty analyze extant environmental communication course syllabi to gain perspective on what is happening in a range of classrooms. Analyzing objectives, assignments, required readings, and subtopics, they find the courses under study are consistent with the focus of research in the field as well as the diversity in the range of topics and multidisciplinary orientations. The authors maintain environmental communication pedagogy should remain open and avoid becoming overly standardized or limited. Milstein, Alhinai, Castro, Griego, Hoffmann, Parks, Siebert, and Thomas explore ways environmental communication pedagogy offers an opportunity to turn conventional learning inside-out and connect learners' *inner* queries and passions to course concepts to extend knowledge building and engagement *outside* classroom walls and institutional boundaries. As they explain in **Breathing life into learning: ecocultural pedagogy and the inside-out classroom**, the "inside-out classroom" and its interface with the ecocultural orientation of environmental communication, emphasizes deconstructing separations, such as those between humanity and nature or classroom and outside world, and on nurturing change agents. The chapter's writing format mirrors the co-creative pedagogical approach, with learner voices bursting throughout and extending beyond the teacher's hopeful seeds of intention.

Part II, Diverse practices in teaching environmental communication, is broader in reach but more specific in engagement, exploring particular teaching practices. In **The role of social constructionism as a reflexive tool in environmental communication education**, Hallgren examines ways social constructionism and symbolic interactionism contribute to the development of reflective and meta-reflective competence in environmental communication education. He describes how this perspective has been successfully implemented in the Master's program in environmental communication and management at the Swedish University of Agricultural Science in Uppsala. Phillips engages questions of intercultural literacy and experiential learning in **"Deep impressions": the promise and possibilities of ecocultural experiential learning** as he investigates the efficacy of outdoors experiential learning in shaping learner attitudes toward a second language and the natural world. In doing so, he documents an experiential assignment he conducted with English learners from various countries during a four-day field study in Yellowstone National Park, braiding together theoretical foundations in experiential learning, (eco)cultural exchange, and language learning to examine ways accessing these in one's pedagogy can foster connections with place and one another. Pedelty and Hamilton grapple with the difficulties of adopting performance and experiential approaches to teaching environmental communication when one takes a face-to-face course online in **Further afield: performance pedagogy, fieldwork, and distance learning in environmental communication courses**. Whereas the authors argue classroom-bound teaching can keep students mostly passive and this trait can be emphasized with a transition to online teaching (a particularly egregious problem in environmental communication education), combining outdoor fieldwork and performance (e.g., an interpretive talk in situ) may offer solutions. In **Arts-based research in the pedagogy of environmental communication**, Takach discusses the nascent but natural alliance arising between arts-based forms of presentation and environmental communication pedagogy. Highlighting key goals, benefits, and challenges of arts-based research and pedagogy, he provides the example of his eco-documentary about the economic and environmental aspects of tar sands exploitation in Alberta, Canada, and explains how he implemented this arts-based approach to structure the capstone projects of his senior students. Lopez explores ways visual literacy can be used in teaching environmental communication in **Developing visual literacy skills in environmental communication**. He draws on his ecomedia literacy framework, which incorporates four approaches to analyzing and engaging media: worldview (the impact of media on "mindfulness," especially our perception of space, place, and time), environment (the material and physiological characteristics of media, in particular communications technology), political economy (political and economic conditions shaping communications), and culture (the symbolic realm shaped by media). Completing this part with **Designing a distance education course in environmental journalism**, Mocatta addresses the trend toward flexible, self-directed, and life-long learning that is increasingly manifesting in the form of online or distance education. Using an online environmental journalism course she created as a case study, she addresses instructional design and assesses ways aspects of distance study may be more or less effective for the goals of preparing professional environmental communicators.

Whereas the previous parts address ways learning and praxis can help nurture agents of change, **Part III, Transformative practice: nurturing change agents**, is centrally driven by this goal. This part showcases pedagogical practices especially designed to engage learners in local and far-reaching environmental issues to help them invigorate their voice and their efficacy. Typhina opens with **Changing *our* environmental future: student praxis through community inquiry**, and presents four strategies for curriculum design that help students become change agents through exploration and engagement with peers and community organizations. Typhina explains how to integrate rhetorical and behavioral environmental communication theories into curriculum through role-playing, activities, and community projects, all of which help students understand how they can use their unique personalities and understanding to bring change to their communities. In **Storytelling as action**, Pileggi and Morgan consider the importance of storytelling in environmental communication in facilitating a "legitimate theater for practical action." The authors describe an exercise, based on the work of Marshall Ganz, that progresses from a *story of self* (to key into student observations) to a *story of us* (to identify key values at play in an issue at hand in one's community) to a *story of now* (to create campaign strategies to address that issue), and provide an example of a class at a university in California creating a campus-wide campaign of awareness and advocacy to address effects of, and behavioral responses to, drought. Walker's ***Insider Windows* in Nepal: a critical pedagogy for environmental communication** reviews ways critical pedagogy, with its foundations in social and ecological justice, can inform environmental communication pedagogy and describes its application to a community-based consciousness-raising project. The community pedagogy project in Nepal engaged youth participants living at the margins of global society in video making to generate subjective participatory environmental and cultural meanings, develop critical awareness, and invite discussion and debate. Kannengießer introduces a hands-on way to explore materiality in **Repair cafés: reflecting on materiality and consumption in environmental communication**. She uses events in which people meet to repair their everyday objects as pedagogical opportunities to explore material aspects of environmental communication by critically looking at production, consumption, and disposal of media technologies (e.g., computers, phones, etc.). To this end, she developed a pedagogy of engagement which incites her students to learn about political, economic, and ecological implications of product life cycles; facilitates students in organizing and operating two repair cafés; and helps students meaningfully sense the materiality of their product use patterns as they hone their skills of persuasion when interacting with the public attending their cafés, promote and practice sustainability practices specific to their learning, and experience being change agents. In **Cultivating pride: environmental engagement and capacity building in the UTEP–Rare partnership**, Tarin,

Upton, Sowards, and Yang describe a partnership between the Department of Communication at the University of Texas El Paso (UTEP) and Rare, an international environmental NGO informed by an outcomes-focused intervention-based pedagogy invested in building co-constitutive capacity to create community-driven environmental messaging. This praxis pedagogy orientation is aimed at enabling local communities to become agents of change and serves as an exemplar for organizations and practitioners interested in fostering substantive and localized environmental engagement.

Finally, **Part IV, Environmental communication pedagogy and practice toolbox,** provides a practical repertoire of activities and exercises educators can use in their own teaching. Activities provide opportunities to fine-tune familiar in-class and out-of-class exercises and to consider ways one might incorporate innovations to suit pedagogical aims. The contributors are active environmental communication instructors who have effectively used these particular tools in a variety of settings; they offer overviews, instructions, learning objectives, assessment ideas, and reflections. The collection of teaching gifts include: Freeman's **Writing a moral vision statement**, which helps students positively frame what they are *for creating* in the long term and not just what they are against in the short term; Milstein and Griego's **Environmental privilege walk: unpacking the invisible knapsack**, an in-class activity that helps learners recognize privilege in their everyday experiences and learn about environmental justice in an embodied way; Plec's, **An experiential approach to environmental communication**, which presents a progressive series of learning activities, including tracking consumption, enacting a low impact day, and doing service learning, providing students with opportunities to identify their practices as environmental actions that have material consequences; Dionne's **Greening epideictic speech**, which tasks students with developing and delivering an original ceremonial address that considers humanity's entanglement within the more-than-human world and creates conditions to experience a form of empathetic unsettlement; McGreavy, Gottschalk Druschke, Sprain, Thompson, and Lindenfeld's **Praxis-based environmental communication training: innovative activities for problem-solving**, which offers several individual exercises that can be sequenced for classroom and community-based contexts to bring diverse perspectives to understanding environmental changes and build core capacities to solve complex environmental problems together; Stroud's **Image(ination) and motivation: mapping environmental positions in visual media** which presents techniques to help students gain insight into the overall efficacy of visual media to convey environmental exigencies and interrogate the ethical potential of such messages; Lopez's **Using infographics to put environmental communication into practice** assignment, in which students learn to create the powerful and increasingly pervasive visual forms of environmental communication and put into practice a variety of concepts; Freeman's **Analyzing news media's incorporation of animal and eco perspectives** assignment, which helps students discover for themselves the extent to which environmental and animal protection issues are addressed in mainstream news media and make grounded critiques about media agenda setting, news construction, and framing; Harrison's **Newschart assignment**, which relatedly helps students develop more sophisticated analyses of media reporting on environmental issues and learn how media environmental representations relate to cultural, social, public, political, and economic affairs; Siebert's **Speaking for/to/as nature** in-class activity, which puts learners in interactive roles of other elements of the more-than-human world to very personally highlight ecocultural meanings, rhetorical uses, and associations; Valencia's **Creating emotional proximity with the environment**, an in-class exercise in which students create a mental map of places where they have been happy and discuss environmental problems that could ruin that happiness, leading to creating environmental journalism that connect their own experiences in ways that are implicative and motivating; Hadl's **Growing up with animals (on screens)**, a group activity in which participants log their most memorable media experiences, screen time, and outdoor time during childhood to explore together how these might relate to their ecological identities and values; and Cruger's **The student-run collaborative environmental communication blog**, a semester-long collaborative undertaking that allows students to build from their experience as consumers and critics of mediated environmental messages to become environmental messages creators in a public sphere.

Every contributor brings a novel approach to understanding how we come to learn what we know and to what is possible in the learning. In doing so, they launch a timely conversation about the pedagogy and practice of environmental communication. We hope readers find this conversation and its contributions beneficial to their own work in engaging ecological wakefulness in their teaching and praxis.

References

Corbett, J. B. (2006) *Communicating nature: How we create and understand environmental messages*, Island Publishing, Washington, DC.

Cox, R. (2007) "Nature's 'crisis disciplines:' Does environmental communication have an ethical duty?," *Environmental Communication: A Journal of Culture and Nature*, 1: 5–20.

Cox, R. and Pezzullo, P. (2016) *Environmental communication in the public sphere* (4th edn.), Sage Publications, Thousand Oaks.

Hansen, A. (2010) *Environment, media, and communication* (Routledge introductions to environment and society texts), Routledge, London.

Hendry, J. (2010) *Communication and the natural world*, Strata Publishing, State College, PA.

Milstein, T. (2009) "Environmental communication theories," in Littlejohn, S. and Foss, K. (eds.), *Encyclopedia of communication theory*, Sage, Thousand Oaks, pp. 344–9.

Stibbe, A. (2015) *Ecolinguistics: Language, ecology, and stories we live by*, Routledge, London.

Part I

(Re)conceptualizing the environmental communication classroom

These chapters focus specifically on ways educators are orienting the environmental communication classroom found in many colleges and universities and how they can (re)orient and (re)locate the classroom.

1 Negotiating knowledge and advocating change

Teaching for transformative practice

Garrett Stack and Linda Flower

Environmental rhetoric is a transformative force working change on self and society, even within the special context of college education. Practicing environmental rhetoric can lead individual students to explore, challenge, and locate their own understanding and values within that of a larger discourse community, including their peers. It can also draw them into translating that negotiated understanding into a public act. Students prepared to speak as environmental advocates are often ideally positioned to enter local publics, those spaces where people may face a shared problem—but do so from differing perspectives, concerns, assumptions or values. It is these two sorts of rhetorical work, negotiating self-knowledge and advocating social change, that can help environmental pedagogy realize its potential as a transformative force.

We find it helpful to think of this transformative model of environmental communication in two ways: first, as a set of challenging roles it assigns to student communicators and, second, as a repertoire of powerful, but frequently demanding, communication practices. That is, the student as a *knowledge-negotiator* must analyze, synthesize, and respond to a wide variety of arguments about the environment. As a *public advocate*, that student must not only make a meaningful contribution to an issue or need, but also draw a public itself into the adaptive work of change, addressing conflict and reframing a shared problem. This chapter will draw on our experience adapting this approach to environmentally focused communication courses as well as advanced seminars in environmental rhetoric. Across these differences, the common thread is an outward-bound focus on rhetorical engagement in a larger community and on an experimental, strategic, and reflective stance on one's own rhetorical action.

To work towards transformation, this chapter outlines two roles for students to inhabit. As a *knowledge-negotiator* in environmental discussions, students work to understand and respond to specific case-based environmental issues in ways that privilege analysis, composition, and a problem-solving process. Learning to formulate reasoned opinions within a diverse set of existing beliefs and arguments, students are also beginning to enter the discourse of environmentalism and its complex arguments.

A second student role, which we call the *public advocate*, approaches environmental rhetoric in order to engage in its public discourse and debate: to enter the field and learn its history, discourse, and values, which shape what people do. Reading and comparing the communications of naturalists, scientists, policy makers, and activists lets the student look at how competing discourses define one's relationship to the natural world, frame environmental problems, or argue for public action. Turning analysis into rhetorical strategy prepares the student *public advocate* to make a genuine contribution as a public communicator.

When the two approaches are treated as a transformative process, environmental communication offers fertile ground to plant seeds for students-as-negotiators to develop responsible communicative practices regardless of—or in addition to—personal opinion. A deeper study of the subtexts surrounding current political and social debates leads students-as-advocates to a more focused study of environmental rhetoric and encourages students to find stakes in this critical set of public choices and contribute meaningfully beyond the classroom.

The knowledge-negotiator

In their article on pedagogy and sustainability, Sprain and Timpson (2012, 342) argue that:

> students typically arrive at college thinking in very dichotomous ways, i.e., ideas are right or wrong, that reality is black or white. When these students begin to hear divergent opinions in college, when they begin to see the complexity in the issues they study, they open up to the "greyness" in life.

Positioning students as *knowledge-negotiators* requires, first and foremost, the construction and support of responsible discourse practices that privilege a wide variety of environmentally focused communications. This positioning helps students to see themselves as not only negotiating the *range* of topics and opinions that exist about various environmental topics, but also negotiating effective ways of communicating *about* these topics themselves. As Gregory Clark (2002) argues, many students enter communication courses with a grounding in the five-paragraph theme, but we may be doing students a disservice by failing to teach a more rigorous means of communicating that would encourage not only more awareness of context and community, but would also provide them with skills that are transferable across other departments within the university, and beyond.

As an alternative, Clarke urges breaking away from the perception that writing exists merely inside of genres that act as containers for text-as-object and suggests instead focusing on writing and classroom discussion as participating in "public discourse" (ibid., 111). One such way of framing the notion of public discourse within the environmental communication (EC) classroom is to focus on case studies. As Sprain and Timpson (2012, 533) suggest, case studies offer students a way to access real-world situations and apply rigorous analysis to develop creative solutions, and argue that "case studies are a productive approach to teaching *about* sustainability and teaching *for* sustainability." According to the authors, this pedagogical tool has interdisciplinary (and transdisciplinary) applications that allow for more productive collaboration within and beyond academia[1] (ibid., 534). Mindfulness of the importance of public discourse is a necessary pedagogical practice for EC, and case studies are indeed an excellent way to approach public discourse within the classroom with an eye towards inter(trans)disciplinary communication. The model we utilize for this course is developed around the presentation, discussion, and written response to increasingly complicated case studies, which culminate in a final paper in response to a particular EC issue.

However, our course diverges from the Sprain and Timpson model in that what we suggest is not the abandonment of teaching *about* sustainability *for* sustainability—and/or other important environmental issues—but to delay the impetus to teach *for* a particular environmental agenda until students have had the time to reach their own conclusions about the issues under discussion. The *knowledge-negotiator* practice adds a way of re-envisioning the EC classroom as a site for instruction and as a community wherein communicative exchange occurs to help students begin to understand and negotiate environmental public discourse. Our model for such a classroom conception comes from Jurgen Habermas's notion of the European "salon" at the turn of the eighteenth century. For Habermas (1989: 33), the salon represented a shift in what was considered "public" from the buildings of state to locations where nobility might join the bourgeois stratum to engage in lively discussions in which "the mind was no longer in the service of the patron; 'opinion' became emancipated from the bonds of economic dependence." The concept of the salon was a novel one at the time because one's social stratum (the prince or the watchmaker) did not matter, and there were no consequences for disagreement; all ideas were given equal consideration. This concept may feel familiar to any instructor who opens a topic for discussion and watches as students, from a variety of socio-economic backgrounds, all have an opportunity to voice their opinion.

If this feels redundant (why must we label a model that is already in common practice?), it is important to note an additional function the salon played in its time: it was a site not only for exchanging ideas, but also for vetting them. As Habermas contends:

> There was scarcely a great writer in the eighteenth century who would not have first submitted his essential ideas for discussion in such discourse, in lectures before the academies and especially in the salons. The salon held the monopoly of first publication: a new work, even a musical one, had to legitimate itself first in this forum.
>
> (Ibid., 34)

The salon then functioned as a kind of peer review, where all ideas, the good and bad, were discussed and weighed equally before being dismissed, reimagined, or approved as sound and novel. Envisioning the EC classroom as a salon, we begin to see it taking on a new shape as an ideal space for public deliberation about environmental case studies: ideas are circulated and vetted throughout the semester through course discussions, readings, and assignments, thus allowing students to negotiate a wide variety of arguments and opinions to give them ample time to reach conclusions of their own.

Adopting the notion of the classroom-as-salon means first privileging discussion of environmental case studies, the free exchange of ideas responding to this corpus, the mutual vetting process, and then positioning the students within the salon as negotiators whose responsibility it is to make sense of the texts within the context of this community. Simultaneously, students are building the communication skills needed to express their opinions beyond the classroom, thus encouraging the interdisciplinarity of course content and communicative practices. However, it is up to the instructor to design a set of practices that are based on case studies to attain flow across a variety of environmental issues and provide real-world environmental situations. First, by introducing basic arguments about the environment, instructors may simultaneously begin a discussion of environmental issues while guiding students through the basics of argumentative structure.[2] As the course progresses, adding increasingly complex case studies[3] to the corpus helps students to become more aware of the nuanced opinions that exist in the debates over current environmental issues, the history of the environmental movement, and the socio-political context within which these arguments exist, while at the same time aiding students to analyze, interpret, and synthesize these arguments as negotiators. By situating these tasks within the framework of environmental case studies, and by engaging students in open debates throughout the semester concerning these environmental problems, the students will feel engaged as negotiators in the subject material while building their rhetorical skills. Ideally, they will feel ready and willing to form their own opinions and share them (via a circulated paper based on a case of their own choosing[4]) with their peers, thus participating in the vetting process of the salon model.

The second positioning required within the *knowledge-negotiator* salon involves the role of the instructor. Taking into account our passionate attachments and intellectual histories,[5] how can EC instructors best position students as *knowledge-negotiators* within a salon—which privileges open debate and the free exchange of ideas—without sacrificing the instructors' own opinions, knowledge, and beliefs concerning environmental issues? We suggest instructors envision themselves as hosts, similar to ones found in political debates, to allow the introduction of ideas and the curating of diverse opinions without forcing students to think one way or another about particular environmental problems. Another way to consider the concept of the instructor/salon relationship is via the "classroom as ecosystem." Under this notion, "[instructors] conceive of educational settings like classrooms as constituting systems akin to biological ecosystems, and propose that research focus on the relationships among the various elements present in the classroom environment" (Guerrettaz and Johnston 2013, 782). As humans are currently demonstrating, any ecosystem can be destroyed when an overconsumptive element is introduced, and an instructor who overburdens students with his or her own opinions may be unbalancing the delicate scales of the salon negotiation process.

The debate over advocacy in the classroom is a rich one, and a number of opinions have been produced about its role in the educational process.[6] We are not asking instructors to discard their own passionate attachments to environmentalism but ask them instead to provide a corpus of case studies representing all sides of an argument so students might reach their own conclusions through the open discussion of the salon model. Here, instructors refrain from voicing their opinions directly (i.e., teaching *for* sustainability), and ask questions that generate further discussion. So instead of explaining the problems inherent in an oil-dependent energy economy, an EC instructor might provide a case study of a college's attempts to divest itself of oil-generated income, highlighting both the pros and cons of the process.[7] We are, in no way, supporting the view that a classroom free of an instructor's advocacy is the best model; in fact, in the next section of this chapter we suggest transitioning the instructor role to one of environmental advocacy. However, for this particular model to work towards transfer, it is our belief that allowing students to engage properly in the salon community requires a free and open exchange of ideas that considers all opinions equally concerning each environmental issue.

As a result of this conscious choice of roles, the EC instructor allows the salon to function as a prerequisite to environmental advocacy, and in envisioning this course as the first step in a transformative process, the salon allows for the gradual construction of an advocatory opinion[8] within the relative safety of the classroom-as-community. Positioning students within the salon as *knowledge-negotiators* serves a variety of functions:

- First and foremost, the salon model allows for open and engaging discussions that introduce and build an environmental corpus through which students negotiate a wide variety of opinions on a range of problems.
- Simultaneously, it requires students to practice active rhetorical analysis and to strengthen their communicative abilities so that they may function well as communicators in a multiplicity of situations within and beyond the university.
- And, finally, it encourages students to construct their own beliefs and opinions that undergo vetting within the salon community so that they may feel empowered to take the next step in the transformative process: the position of environmental advocate.

The transformative public advocate

Advocating for the environment must go beyond simply taking a personal stand or arguing for one's position on a proposition. The *transformative public advocate* we are describing must draw others, a local public, into enlarged understanding. As the research on such communication suggests, speaking for the environment means understanding not only the expectations and affordances of multiple discourses, but also the (competing) arguments and expectations of the public communicators hope to call into adaptive work. Killingsworth and Palmer's (2012) thought-provoking history of "Ecospeak" shows how environmental rhetoric speaks through a remarkable array of discourses, from John Muir's elegies for doomed valleys to Greenpeace's daredevils crafting publicity. Equally remarkable is the range of roles and strategies an individual such as Aldo Leopold may require as the Sand Hill farmer/U.S. Forestry official/professor of game management/public figure crosses from the discourse of nature observation to the reflective persuasive essay, professional report, or embattled political argument.

All these accounts raise questions of rhetorical choice. Just how did the Inuit use techniques of association and dissociation to convince New Yorkers not to buy Canadian water that would have flooded native hunting grounds (Cooren and Taylor 2000)? Or why did Muir's poetic pleas for Hetch Hetchy fail in the face of the cruder economic arguments of Gifford Pinchot or the clever pamphleteering of Congress before a vote (Oravec 1996)? And, of course, how did that loss become the springboard for a public awakening and the preservation ethic to come?[9]

These are the sorts of problem that drive the pedagogy sketched here. One of the first steps in becoming a *transformative public advocate* is to become an astute reader of rhetorical strategy in others' texts. Each primary text is read not only for itself, but also in combination with a critical commentary that demonstrates a distinctive method of analysis. Class discussion then offers a chance to test-drive a different critical concept from argument theory, decision research, or rhetorical analysis to see what it can reveal. When students *use* scholarly ideas to figure out the rhetorical work a text is doing, they see the power of different analytical options. Is the art of building an argument here best understood with Toulmin's (2003) breakdown of "data/warrant/claim," Burke's (1969) account of the power of "identification," or Aristotle's (1991) venerable trio of 'ethos, pathos, and logos"? These "available means of persuasion" constitute a rhetorical techne[10] drawn from multiple disciplines—a set of strategic tools—for not only unpacking the arguments of others, but also building one's own.

Students need to realize the tradeoffs different methods offer to make effective advocates. Critical discourse analysis, for example, positions *public advocates* within the framework of an "explicit sociopolitical stance," that, as van Dijk (1993, 252) argues, spells out "point of view, perspective, principles and aims … [that are] admittedly and ultimately political. Their hope, if occasionally illusory, is change through critical understanding." Well-designed to capture highly verbal speech acts and argumentation, it can also reveal non-verbal elements, such as the semiotics and visuals that figure in Gore's *An Inconvenient Truth* (Wodak and Meyer 2001, 2). Given its own premises, it can ruthlessly unpack hidden assumptions in the debates about environmentalism found in essays, newspaper op-eds, grass-roots pamphlets, political speeches, and film. Equally valuable, however, is Waddell's (1998) account of how writers attempt to actually "re-constitute" or reshape the values of an audience based on the art of drawing a public into "co-creating" a more comprehensive paradigm. And shifting the focus to audience means we must consider arguments and outcomes. In still other cases, the research on communicating risk from decision sciences could shape our rhetorical art in terms of the critical kinds of knowledge people need for choice and the psychology of response (Morgan et al., 1992). Comparing and testing out alternative rhetorical tools, with an eye to productive, transformative work is an important step toward effective advocacy.

These ideas become clearer in the context of the following environmental rhetoric seminar, in which students analyze an environmental controversy to propose a more inclusive argument strategy illustrated with a prototype text. Students first use their analytical skills to explore the public discourse that surrounds an issue they care about and to identify a "client" whose public work they wish to support.[11] Positioned as a pro bono consultant, they will offer a researched analysis of the client's rhetorical problem in the context of the critical arguments, assumptions, and approaches already in circulation.[12] This initial rhetorical problem analysis will then be paired with a recommendation supported by a prototype of a public document (a text, brochure, sample web page), justified by an analysis of its underlying rhetorical choices.

Learning how to mine texts and translate research into an expanding repertoire of strategies lays a foundation for the transition from the student-as-negotiator within the salon to student-as-rhetorical leader beyond the classroom. Ronald Heifetz's (1994) influential study, *Leadership Without Easy Answers*, identifies success with the savvy and skill it takes to translate what might be relegated to the status of a "technical problem" (e.g., ways to reduce the effects of carbon emissions) into "adaptive problems" that call everyone involved into facing the need to change. The role of the leader—who must call disparate, even opposed stakeholders into adaptive problem-solving—becomes intensely rhetorical. And a leader/advocate who wants to call a public to participate in John Dewey's (1927) "great community" (where democratic deliberation is focused on *consequences* for all) must be able to build an argument that models adaptive thinking and communal problem-solving.

The following student project shows one way the *public advocate* stance let a student integrate learning with a genuine contribution to environmental understanding. In "TroutReach: A Narrative Action Plan," the writer's proposal argued that the public face of the Pennsylvania Council of Trout Unlimited had both a "communication problem" and a missed opportunity. In public debates the fishing organization was being misrepresented as "sportsmen," a position, which he felt, did not fully represent the environmental ethos that was an underactivated part of the group's identity and mission statement. Drawing on the techne of "rhetorical narrative," the writer decided he could take a rhetorical leadership role by helping the Council craft a new way for fishermen to embrace this environmental ethos through public narrative. Building on that common narrative trope that unites fishermen (the "fish tale"), the "TroutReach Narrative Action Plan" offered a step-by-step guide inviting fisherman to turn these common, verbally shared fish tales into written stories that would help them reframe their role as not only sportsmen but also environmentally conscious citizens working for public good. In addition, *TroutReach* provided guidelines for the Council members to help get these stories published in broader public forums. The student's final project combined this practical guide with a consulting report analysis for the Council of both the existing public discussion and a rhetorically sophisticated justification for the rhetorical narrative approach he was recommending. Though the student never received a response from the local council, the work was submitted, thus validating the "publicness" of the project; it may yet be used some day for the betterment of local watersheds.

The goals and potential value behind such a project reflect our reading of public sphere theory. Habermas's (1989, 27) influential study of the rise and transformation of the public sphere is in part a history of an emergent form of "political confrontation … without historical precedent: people's public use of their reason." But Habermas's account is also a normative model, based on the ideal of a critical–rational form of deliberation, in which we merely "bracket" difference and argue not from interest but general principles. His story ends with the decline from this ideal into our contemporary media-based public sphere co-opted by corporate and political interest.

However, if we look at the dynamics of "actually existing democracy" as Fraser (1990, 27) suggests (versus assuming a normative model) we will see some problems with the ideal of critical rationality that would affect our pedagogy. First, in "stratified societies, public work does not happen in 'a' public sphere, but in 'a plurality of competing publics'" (ibid., 66); that is, there are lots of "publics" out there actively competing for us to see the issue their way. Moreover, controversy and social change are typically initiated in grassroots groups and local publics, not official deliberative forums (Crick and Gabriel 2010). And counter publics, which have the power to call a new public into being and to circulate a new discourse, are often the force that alters the way we talk and think (Warner 2005). In a final contrast to critical rationality and the elite discourses of formal analysis, expertise, or policy, democracy speaks with vernacular voices offering practical reasons, doing so in the language of narrative (Young 1997, 131) that privileges difference and the situated knowledge of others, and in contrast to the ideal of competition among self-interests values the generative possibility of "collective problem-solving" (ibid., 401).[13]

The "TroutReach Plan" was, in essence, an attempt to energize a silent constituency and bring it into a public debate. It did so first, not by arguing from broad principles, but by offering fishermen practical good reasons for entering public discourse, and then by turning the potential of their own situated, performative knowledge into an artful rhetoric that might draw others into collective problem-solving around threatened local watersheds and wildlife health. Obviously this style of advocacy differs from the aggressively argumentative attack that corruption and greed in the world of environmental politics do indeed merit. Although it may work hand-in-hand with the adversarial argument that energizes marches, pressure groups, and funders, *advocacy for deliberative engagement* has a different role to play. Moreover, it is a rhetorically difficult one that justifies equally sophisticated skills. In terms of a pedagogy designed to aid in transforming the student role from *knowledge negotiator* to *public advocate*, we read this body of theory as raising three challenges:

- Instructors must help *call a new public* into being around a shared interest—to induce people to recognize themselves as part of this expanding public connected by a moving environmental concern.
- They draw this network of people (with their differences) into the work of *framing and reframing the problem* in an act of collective problem-solving; that is, they help scaffold a rhetorical engagement—which recognizes the competing values and rival hypotheses of multiple stakeholders—while building one's best case for wise environmental choice.
- Finally, instructors help students translate understanding into a call to *personal and public action*.

This process may seem quite doable in a congenial community of fishermen. But what about environmental issues that are already locked in a polarizing debate where "ecospeak" is hurled between opponents from entrenched positions? For instance, 2014 saw an inflamed debate labeled by some as the "war on wolves" when state governments allowed unlimited killing of wolves by any means. Environmental websites went on emergency attack. But where does the long-term solution lie in this struggle? Is there a common ground, as this student writer suggests?:

> Many humans have feared and hated the "big bad wolf" for centuries.... There's no question that wolves can be dangerous. Wolves are wild animals that are not our friends or pets: they're simply fellow inhabitants of the same land as us.

To illustrate this stance, the student's prototype was a small brochure titled "Coexistence Compromise: Learning to Live Alongside Wolves." On the cover, the sun-worn profile of a Western rancher faces left, above the profile of an equally thoughtful wolf looking to the left. Written for the non-profit Defenders of Wildlife, it speaks to competing perspectives, outlining a compromise based on the benefits the wolf population has—not just for the ecosystem, but also for hunters, for pest control, and for other species. For example, one section titled "Fact vs. Fiction: Wolf Myths Debunked" speaks directly to ranchers: "Fiction: wolves kill livestock whenever they get a chance. Fact: They only encroach on ranches when they're forced out of their natural habitats by human development or reduced wild prey." Acknowledging contexts where protection is necessary, another section reviews "non-lethal control methods" as a viable alternative to extermination and an example of the non-profit's position, framed as "Let's Work Together."

The logic of what is in fact a deeply "pro-wolf" argument becomes even more apparent when we compare it to an earlier draft. The writer's first impulse was to elicit empathy used photos of smiling researchers and wildlife advocates consorting with people-friendly wolves: an image likely to resonate with the "converted," but not with ranchers envisioning dead sheep. The rhetorical challenge for this writer was to create an invitation that would draw a wider public to see this problem as *their* problem—whatever their first priorities. In this course, the prototype brochure was *not* the end goal: explaining and justifying his rhetorical strategy was. In renaming the issue as the challenge of "co-existence," he reframed the problem from a battle between animal killers and animal lovers, or money and ethics, into a choice to acknowledge multiple options, then figure out how to act wisely on values an entire community could hold.[14] Reframing the conflict as an issue clouded in *misconceptions* that call for active listening, he reconceptualized the advocate's "war on wolves" into an adaptive problem—a challenge to "co-existence" that called his public to the difficult work of finding better alternatives.

The problem a student defines and the strategic approach he or she argues for is also a rhetorical choice. For instance, when proponents of a new wind farm on the Chesapeake Bay needed to convince residents of its local and environmental value and reduce fears of "ruined views and land values," one student chose to respond through identification. She used interviews with actual stakeholders to evoke the presence of "neighbors" sharing positive experiences and pictured them in a brochure for local distribution. When she engaged in out-of-class "salon" discussions with another student (who was drawn to the same farm, but with a broader activist interest in the technology), they discovered room for strikingly different representations of the rhetorical problem. He saw the wind farm's website as a potential model for other, less-well-funded groups and took on the delicate, diplomatic job of showing the downsides of their current highly informational approach. He argued for a rhetoric of engagement and modeling techniques to create a reflective voice.

Sometimes the real problem is actually calling a public into being and initiating deliberative discussion through what Warner calls "circulation." An EC student who felt a personal identification with North Dakota farmers also knew them to be suspicious of urban "environmentalists" and so designed her work for AgUnited, a local farmer's cooperative, and monthly *Farm Forum*'s Green Sheet. The text she created is framed around pragmatic questions farmers have about the cost, benefits and trade-offs of using widely promoted neonicotinoid treatments on seeds—a chemical that is linked in a growing body of research to the destruction of honeybees, a critical crop pollinator. The farmers might ignore a call to "Save the Bees," but might find the facts about the actual trade-offs—and the real effects on yield—revealing. Moreover, her text frames the decision as a topic for discussion *among farmers* in their own *local publics*. It calls together a new public of people who care intensely about their land.

The final project in this seminar is not a flashy, finished brochure or website but a prototype or mock-up of a public document, accompanied by the two-part consulting report. Having examined the research, arguments, perspectives, and propaganda surrounding the issue in part one, part two must build a persuasive case for the student's proposed rhetorical response, analyzing the attached prototype of a public document in terms of the options, strategies, and rhetorical choices it represents. As an educational project, the focus here is not on document design but on thinking through rhetorical options in the face of a genuine need. As a professional effort modeled on a response to an RFP and a portfolio piece, it lets some students build a network of real contacts in an area, while for others it is a laboratory for learning to put analytical tools and techne into socially engaged practice. In either case, the real challenge is for students to see their work as a transformative public act.

Conclusion

We are part of a growing effort to link environmental rhetoric and communication with work in the broader public sphere. In her discussion of ecopedagogy, Rhonda Davis (2013, 81) perhaps best captures the goal of publicly-minded environmental communication courses when she argues:

> In order to effect socially responsible change, which many composition scholars believe to be an integral goal of teaching, it is critical to embrace this ecological concept in teaching composition and rhetoric, as well as to engage in public discourse.

The goal of effecting socially responsible change is indeed doable, and the burgeoning field of ecopedagogy is a testament to the successes this type of rhetorical training can produce.

This chapter adds a framework for inviting EC students into the role of knowledge makers and public performers through a transformative stance built on transferable skills and a deepened knowledge of both rhetoric and environmentalism. Making the transition from *knowledge-negotiators* to *transformative public advocates* combines rhetorical communications skills with socially responsible change. Engaging students in the free exchange of ideas within a specified community, the salon offers a risk-free method to not only hone analytical and communicative skills but also to vet and negotiate opinions about specific environmental cases. Turning to real communities in need of adaptive problem-solving, students can apply their theoretical understanding and rhetorical art to transformative public work. The "salon-effect" and *knowledge negotiation* are intimately linked to *transformative public advocacy*. Both are steps in a process that leads from open-minded exploration to responsible, community-based engagement.

Notes

1 Such as with community-based projects like those undertaken as a *transformative public advocate* as described in Section III.

2 Aldo Leopold's essay "Thinking Like a Mountain" (Leopold and Schwartz 1970, 129–33) is an excellent example of a narrative-based argument structure that also introduces the basic components of land and game management, ecosystem health, and ecology.

3 Arguments over fossil fuel divestments, such as the work found at www.gofossilfree.org, is an excellent example of a more complex case study that can be applied to many local campuses, cities, and/or nations.
4 Topical freedom for the final paper (under the umbrella of environmentalism) tends to produce delightful variety, where case studies range from collegiate cell phone use to ecoterrorism.
5 See Ede (2004) for further discussion of "passionate attachments" and "intellectual histories" and their importance in modern communication scholarship.
6 Spacks' (1996) work on classroom advocacy is one such collection of these opinions and a fine source for more information on the subject.
7 Robert Cox (2010, 126) lists a series of successful case studies on community collaboration (126) that might also be useful.
8 See Flower et al. (2000). They discuss rival hypothesis thinking as an inquiry-based perspective that addresses all problems as open questions which have multiple answers and encourages advocacy and rivaling as a means reaching shared conclusions.
9 Some rich sources for such controversies include not only Killingsworth and Palmer (2012), but anthologies edited by Herndl (1996), Coppola and Karis (2000), and Waddell (1998).
10 As Atwill (1998, 48) argues, *techne*, translated as *art*, was associated with "productive" (versus "theoretical") knowledge in rhetoric. And as a "domain of intervention and invention" it is often associated with resisting boundaries and the status quo.
11 "Clients" can range from a campus office convincing/showing students and staff how to "go green," to community groups and local publics called around an issue (doesn't fracking also create jobs?) to non-profits addressing environmental problems.
12 Students here are responsible to the needs of a genuine client but not required to enter into the direct contact and often time-consuming social negotiations of a community project. Designing the project as essentially a response to a request for proposal (RFP) rather than as a community collaboration has some other important educational advantages. Students can set imaginative, personally motivating goals, such as convincing your rural "I love my pick-up [truck]" neighbors to support rural transit. For others, the project is preparation for a summer internship in environmental engineering or a portfolio piece in a career directed toward public policy, decision science, education, or professional writing.
13 For more information on vernacular voices and situated knowledge, see Hauser (1999) and Flower (2008) respectively.
14 Cox (2010, 125) may shed some light on the subject in his discussion of community-based collaboration as opposed to arbitration and mediation.

References

Aristotle (1991) *The rhetoric of Aristotle*, trans. L. Cooper, Appleton-Century Crofts, New York.

Atwill, J. (1998) *Rhetoric reclaimed: Aristotle and the liberal arts tradition*, Cornell University Press, Ithaca.

Burke, K. (1969) *A rhetoric of motives*, University of California Press, Berkeley.

Clark, G. (2002) "Departmental Syllabus: Experience in Writing," in Roen, D., Pantoja, V., Yena, L., Miller, S. K. and Waggoner, E., eds., *Strategies for Teaching First-Year Composition*, National Council of Teachers of English, Urbana, IL, pp. 102–13.

Cooren, F. and Taylor, J. (2000) "Association and Dissociation in an Ecological Controversy," in Coppola, N. and Karis, B., eds., *Technical communication, deliberative rhetoric, and environmental discourse: Connections and directions*, Ablex Publishing, Stanford, pp. 171–90.

Coppola, N. and Karis, B., eds. (2000) *Technical communication, deliberative rhetoric, and environmental discourse: Connections and directions*, Ablex Publishing, Stanford.
Cox, R. (2010) *Environmental communication and the public sphere*, Sage Publications, Thousand Oaks.
Crick, N. and Gabriel, J. (2010) "The conduit between lifeworld and system: Habermas and the rhetoric of public scientific controversies," *Rhetoric Society Quarterly*, 40(3): 201–23.
Davis, R. (2013) "A place for ecopedagogy in community literacy," *Community Literacy Journal*, 7(2): 77–91.
Dewey, J. (1927) *The Public and its problems: An essay in political inquiry*, Swallow Press, Athens.
Ede, L. (2004) *Situating composition*, Southern Illinois University Press, Carbondale.
Flower, L. (1981) *Problem-solving strategies for writing*, Harcourt Brace, New York.
Flower, L. (2008) *Community literacy and the rhetoric of public engagement*, Southern Illinois University Press, Carbondale.
Flower, L., Long, E. and Higgins, L. (2000) *Learning to rival: A literate practice for intercultural inquiry*, Lawrence Erlbaum Associates, Mahwah.
Fraser, N. (1990) "Rethinking the public sphere: a contribution to the critique of actually existing Democracy," *Social Text*, 25/26: 58–60.
Guerrettaz, A. M. and Johnston, B. (2013) "Materials in the classroom ecology," *The Modern Language Journal*, 97(3): 779–96.
Habermas, J. (1989) *The structural transformation of the public sphere: An inquiry into a category of bourgeois society*, MIT Press, Cambridge.
Hauser, G. (1999) *Vernacular voices: The rhetoric of publics and public spheres*, University of South Carolina Press, Columbia.
Heifetz, R. (1994) *Leadership without easy answers*, Harvard University Press, Cambridge.
Herndl, C. (1996). *Green culture: Environmental rhetoric in contemporary America*, University of Wisconsin Press, Madison.
Killingsworth, M. and Palmer, J. (2012) *Ecospeak: Rhetoric and environmental politics in America*, Southern Illinois University Press, Carbondale.
Leopold, A. and Schwartz, C. (1970) *A Sand County almanac: With essays on conservation from Round River*, Ballantine Books, New York.
Morgan, M., Fischhoff, B., Bostrom, A., Laver, L. and Atman, C. (1992) "Communicating risk to the public," *Environmental Science and Technology*, 26(11): 2048–56.
Oravec, C. L. (1996) "To stand outside oneself," in Cantrill, J. and Oravec, C. L., eds., *The symbolic earth: Discourse and our creation of the environment*, University Press of Kentucky, Lexington, pp. 58–75.
Spacks, P. M. (1996) *Advocacy in the classroom*, St. Martin's Press, New York.
Sprain, L. and Timpson, W. M. (2012) "Pedagogy for sustainability science: Case-based approaches for interdisciplinary instruction," *Environmental Communication*, 6(4): 532–50.
Toulmin, S. (2003) *The uses of argument*, Cambridge University Press, Cambridge.
van Dijk, T. A. (1993) "Principles of critical discourse analysis," *Discourse and Society*, 4(2): 249–83.
Waddell, C. (1998) "Perils of a Modern Cassandra," in Waddell, C., ed., *Landmark essays on rhetoric and the environment*, Lawrence Erlbaum Associates, Mahwah, pp. 55–73.
Warner, M. (2005) *Publics and counterpublics*, Zone Books, New York.
Wodak, R. and Meyer, M. (2009) *Methods of critical discourse analysis*, Sage Publications, Los Angeles.
Young, I. (1997) "Difference as a resource for democratic communication," in Bohman, J. and Regh, W., eds., *Deliberative democracy: Essays on reason and politics*, MIT Press, Cambridge, pp. 383–406.

2 Pedagogy as environmental communication

The rhetorical situations of the classroom

Jessica M. Prody[1]

As environmental communication emerged as a discipline, Robert Cox (2007, 15) argued it was a crisis discipline with an obligation to "enhance the ability of society to respond appropriately to environmental signals relevant to the well-being of both human communities and natural biological systems." Although this categorization drew criticism (Killingsworth 2007), recent scholarship suggests that many in the field are committed to the ethical framework Cox outlined. Current scholarship intervenes in environmental debates, clarifies knowledge needed to make policy decisions, provides insights into historical controversies that help illuminate contemporary ones, and raises challenges to debates that exclude marginalized voices (both human and non-human). How this ethical framework functions in environmental communication pedagogy is less clear. Environmental Communication scholars have not readily explored how belonging to a crisis discipline impacts the pedagogy of environmental communication classrooms or how pedagogy is itself environmental communication.

The underlying assumption of this essay is *environmental communication educators are environmental communication practitioners*. As with any good communicator, environmental communication educators must identify their purpose, account for their audience, and adapt to particular constraints imposed on them by the context of their teaching. They must use pedagogical communication that balances their role as good instructors (who teach content, critical thinking, and communication skills) with that of being scholars committed to an ethical framework of improving social deliberation about environmental issues.

To understand the rhetorical skills necessary to produce good environmental communication in classrooms, I draw on three case studies from my own pedagogy. I apply Lloyd Bitzer's (1968) framework of the rhetorical situation to my experiences, identifying the rhetorical situations in which I found myself and detailing how my pedagogical tactics contributed to the success or failure of my communication. I conclude the chapter with a discussion of the significance of identifying educators as environmental communication practitioners and outline a process by which educators might improve the environmental communication of their pedagogy.

The rhetorical situation

Two conclusions follow from the claim that educators of environmental communication are also environmental communication practitioners:

- First, environmental communication is not only taught to increase students' knowledge: it also includes significant persuasive messaging. Environmental communication pedagogy involves:

  a *proving* to our students that environmental issues directly impact their lives and that the language used to discuss the environment has material impacts on the world;
  b *convincing* them that they have various modes of rhetoric (from speeches to artwork to the taking up of space) at their disposal when participating in environmental deliberations; and
  c *persuading* them to see that their discourse has the potential to shape discussions, attitudes, and policy.

- Second, if environmental communication pedagogy is environmental communication, then it should be able to be evaluated and strengthened utilizing existing communicative models. Specifically, Bitzer's (1968) rhetorical situation provides a framework useful for evaluating environmental communication pedagogy and adapting that communication to better meet course aims and student needs.

The rhetorical situation is a productive model to evaluate environmental communication pedagogy as this type of pedagogy meets Bitzer's (1968) definition of rhetoric as pragmatic—it comes into existence for something beyond itself and is used to accomplish specific tasks. It persuades students. It encourages them to identify the impact of environmental issues, to see how environmental communication shapes environmental attitudes and behaviors, and to recognize their capacity and responsibility to participate in environmental discourse. Environmental communication pedagogy also meets Bitzer's definition of rhetoric as situational. Each classroom has different needs (exigencies), different audiences, and different constraints, which cause each environmental communication classroom to require unique pedagogy from the educator.

Bitzer (1968) identified three constituents of the rhetorical situation: exigence, audience, and constraints. The first constituent, exigence, is "an imperfection marked by urgency; it is a defect, an obstacle, something waiting to be done, a thing which is other than it should be" (ibid., 6). The exigences in the environmental communication classroom can be identified as what the students need from the course. Do they need basic knowledge, practical skills, or inspiration? Many environmental communication classrooms contain some combination of these exigencies. The educator determines which exigence needs the most attention, what Bitzer called the "controlling exigence" (ibid., 7), and tailors the content of his or her pedagogical message to meet that primary need.

Similarly, the educator must assess the existing knowledge and commitments of students and identify what can be asked of them as audience members or "mediators of change" (ibid., 7), people capable of fulfilling the rhetoric's purpose—e.g., learning, being persuaded, or engaging in action. Not all students will be in a place where they will transform into environmental advocates in a single course; others may already be participating in marches or town hall meetings. An educator must decide what roles students are willing and able to assume and identify what actions students may be willing to take.

The final constituent—constraints—refers to "persons, events, objects, and relations which are parts of the situation because they have the power to constrain decision and action needed to modify the exigence" (ibid., 8). In an environmental communication classroom these can be something as basic as distracting technology, or something as complicated as lack of knowledge or "green fatigue"—the sense that "individuals see their actions as largely irrelevant when set against the enormity of global climate change" and avoid specific action or even listening to environmental messaging (Wilson 2007, para. 3). Educators must identify which constraints they are facing and talk with students about them, as well as tailor the content of the course to overcome them to meet course goals.

Bitzer's (1968) concept of the rhetorical situation has faced critique, most notably from Richard Vatz (1973), who raised concerns over the assumption of an intrinsic nature of the rhetorical situation. Vatz contended, "Meaning is not discovered in situations, but *created* by rhetors" (ibid., 157). The classroom is an interesting rhetorical situation from this perspective because the instructor and students are communicators who co-create the situation's meaning. Nonetheless, Bitzer and Vatz can be used collaboratively to build productive environmental communication classrooms. Bitzer's model provides a set of questions that helps an instructor identify the communicative needs of student audiences, and Vatz's model reminds us that pedagogical communication must be fluid. An instructor must be willing to suspend his or her initial goals of a course and the communication they inspire if the goals diverge radically from the students' expectations or abilities. In addition, as students' knowledge, interests, and desire for action change, the rhetorical situation is altered, and new exigencies and constraints are introduced which will require an alteration to one's pedagogical communication. Vatz's model reminds us that the rhetorical situation of the classroom is a constant negotiation between instructor and students and requires flexibility. I illustrate the application of the rhetorical situation to environmental communication pedagogy by examining three situations of my teaching, one in which my communication failed to meet the demands of the rhetorical situation, and two others in which I altered my approach to meet the demands.

Assumed knowledge and lack of place

The first environmental communication course I taught was an upper-level special topics course for my department's majors. I devised the course as a survey, covering environmental history, the subfields of environmental communication, and important concepts, such as environmental justice, anthropocentrism, and the nature-culture divide. The course's exigence—what was needed from the course—was to get students to identify how important environmental communication was to shaping their world experiences. I assumed my audience had been exposed to environmental concerns throughout their lives. The constraints I envisioned were similar to those faced in any classroom: short attention spans, distracting technology, and a need to see a direct impact on one's life for course material to have relevancy. I did not believe the content would cause constraints because, to me, it was inherently interesting.

The course's main assignment was a semester-long public education project in which students researched an environmental topic to prepare themselves to teach a public audience about it. This assignment best illustrates how my failure to correctly identify the rhetorical situation negatively impacted the course. Four out of sixteen students chose to focus their projects on littering, recycling, or trash. Although not unimportant in the history of environmental thought, studying trash throughout the semester in a project proposal, background research paper, and audience analysis did not do much to keep these students engaged. My biggest error in evaluating the rhetorical situation I faced was overlooking the largest constraint of my audience—*a lack of knowledge of contemporary environmental issues*. These students chose these projects because trash was the environmental issue about which they felt they had enough knowledge to teach others. This became clear in our discussion of climate change when few could explain what it was or its potential risks. They also didn't inherently see environmental issues as related to their own experiences. Instead, these issues were abstract concerns that affected far-off places. My early environmental communication in the course should have included more content about significant contemporary environmental issues facing society and discussion about the immediate impact of those issues on students' lives. By not beginning in this way, some of my audience members remained unconnected to the course throughout the semester.

The course was not a complete failure, but student responses in evaluations identified the weaknesses in my approach. Common feedback for the course was, "I learned a lot about environmental issues." Although partially positive feedback, it suggests only one goal of the course was met. Students were introduced to content they had not been aware of previously, but only one student's response suggested the primary exigence for the course was achieved. This student wrote, "[This course] has opened my eyes to a new way of thinking and talking about the environment and how it affects virtually every facet of life." This student came to understand the impact that communication can have on shaping our world. I wanted each of my students to come to this conclusion, but structuring the course to emphasize breadth over depth, and assuming more knowledge of environmental issues than the students had, created a course where my communication helped students learn facts, history, and concepts, but failed to help them consider the broader significance of environmental communication.

Discourse in environmental communication classrooms must go beyond informing students about environmental issues and communication concepts. The communication should move students toward action. To get students to this point, however, they need to understand how they and their places are affected by environmental concerns. Rania Masri (2007) argued that one must personalize one's teaching and draw on emotions to help students connect to course material. She encouraged asking deeper questions that guide students toward new ways of thinking. Frankly, some students won't be ready to ask big questions at the beginning of a course, but an educator must help these students connect environmental communication concepts to their lives to prepare them to channel their emotions to ask these questions and move toward participation in creating change. We have to meet our students where they are and tailor our communication to the rhetorical situation they, as audience members, construct. I failed at doing this in the course, and for that reason I am not sure how prepared or motivated my students were to go out into the world and produce environmental communication. In the second iteration of the course, I grounded the course communication in students' lived experiences and found much more success.

Adapting to diverse interests

In my revised course, I narrowed the scope of the class by using food as a lens to study environmental history and environmental communication. I also made major changes to the course assignments that allowed students to better connect with the material and become invested in the environmental communication they produced. These changes allowed me to better meet the needs of the exigence of the course and address the constraint of lack of knowledge or interest in the subject matter that came from the audience.

Following scholars such as Leah Sprain and William Timpson (2012), who argued for the productiveness of case studies in teaching about sustainability, I used food as a lens to communicate about environmental communication, which gave a focus to the course that had been missing previously. Throughout the semester food-themed case studies were used to introduce course concepts and to provide the thread that tied together areas of study. For example, a chapter from Rachel Carson's (1962) *Silent Spring*, which focused on the threats of pesticides to our food supply, allowed the class to discuss environmental history and the rise of new environmentalism. Furthermore, exploring the contemporary food system introduced global climate change and environmental justice issues to our classroom discussions. All students had some investment in food, which allowed them to draw on their personal interests. Economics majors were intrigued by the development of big agribusiness, environmental studies majors were interested in genetically modified organisms, and communication majors were interested in food rituals and cultural constructions of food's meaning. Food case studies allowed students to consider ethical and philosophical questions of environmental concerns—issues often left out of discussions in purely scientific classrooms.

I also used case studies in the design of course assignments. In addition to the public education project, I added a place project that asked students to reflect on the connection between person and place. The place assignment led students to make stronger connections between course material and their lived experiences. Throughout the semester students kept a journal about a place that had shaped their identities. At the end of the semester students produced creative communication to impart the importance of their places to their fellow classmates. One student created a sound installation of an urban park (where he had been taught to play basketball by his uncle and had then taught his younger brother to play) which communicated the passing down of knowledge that marked this place. Another wrote a series of fables to share the life lessons her family cabin had taught her. This assignment encouraged discussions about how we define environment and nature, what causes people to identify with a particular environment, and the role environments play in shaping our relationships with others and the non-human world. It also persuaded students that non-traditional rhetoric could produce powerful arguments.

The impacts of the changes to the course were clear in students' improved public education projects and in course evaluations. Students in this course picked topics for their education projects important to places they belonged. A student from Detroit undertook a letter-writing campaign to Detroit's city council to advocate policies that would encourage urban farming. Another created a lesson plan on invasive species for youth hunting programs in upstate New York. Through these projects, students gained greater understanding of how environmental issues impacted their own lives and gained confidence engaging in environmental communication.

Emphasizing the case study approach in class allowed every member of the student audience to come out of the course with a better understanding of how directly their lives were connected to the environment, how they were shaped by their environmental surroundings, and how their actions could shape the environment. Fulfilling the course exigence was evident in student responses on evaluations. One student wrote,

> As an environmental studies major it was a lot of fun to talk about the environment in a different context. In other environmental courses, we have never discussed how people communicate about the environment or how the environment communicates things.

Another noted, "Environ. Comm. [sic] is a great class to learn about issues, ways to address these issues, and how we can help in our community." The students quoted here came to the class with significant environmental knowledge, but the course kept their attention because it offered a different perspective from which to consider the knowledge they already had and motivated them to practice what they learned in the course in their everyday lives.

The focus of the course also allowed those who came to the class uneducated about environmental issues to identify what their major concerns might be. One student wrote, "I have really enjoyed taking this class, in particular learning about GMOs and eating local foods." Another student was happy they had a better general base of knowledge about environmental issues: "I had never taken an environmental course, so I learned about environmental factors, and seeing how it interacts with communication was really interesting." Environmental communication courses are well situated to reach audience members who, because of an aversion to science, might not naturally be drawn to courses where environmental issues are discussed. Certainly, one needs a basic scientific understanding to ethically participate in environmental conversations, and that can be provided in environmental communication courses; however, detailed scientific understanding of environmental issues is not necessary to understand one's stake in these discussions. In this second iteration of the course I used TED Talks (a genre of lecture available on a YouTube channel and various streaming devices), which explained climate change in terms a layperson could understand, discussed the scientific method, and approached the science of pollution and health effects through social justice concerns. Small changes to my pedagogical communication to address the needs of my audience allowed me to fulfill the needs of the rhetorical situation by acting as an ambassador of environmental science and guiding students toward environmental concerns using questions of ethics and culture.

Adapting to environmentalists

In addition to my environmental communication course on campus, I taught in an off-campus program called The Sustainability Semester. In this program, students live on a farm and take an end-of-the-semester trip to Boston where they compare sustainable rural and urban living. Student schedules include homesteading activities, off-campus community partnerships, and sustainability-related courses in the sciences, humanities, and social sciences. My audience for this course was exposed to sustainability from the time they got out of bed to the time they returned to it. They did not need the basic introduction to environmental issues necessary in the on-campus classroom and were already convinced that environmental issues directly impacted their lives.

The exigence of the situation in this classroom was that students needed guidance about how to implement their program experiences into their existence off the farm. Providing time in the class for students to contemplate their non-traditional educational experience was one way to help students connect the material they were learning to their lives. Another was to use a flexible class structure that allowed students to guide the direction of class time.

Elizabeth Dickinson (2011, 315) argued, "Perhaps environmental education in general should consider avoiding the appropriation of regimented classroom practices so as to let students have less disciplined experiences." I followed this advice by creating course activities that encouraged students to learn outside of the classroom. We took field trips to local trails and spent time standing outside "listening," as we tried to connect to the writings of Leopold and Thoreau (e.g.

Leopold 1949; Thoreau 1908). In one activity, I asked students to go into town and price out the same meal at the grocery store, natural foods store, convenience store, university cafeteria, and McDonald's. We brought the information back to the classroom and discussed the various costs and compared them to the costs of homesteading materials needed to produce the same meal. This allowed us to examine issues of food access and the environmental impacts and costs of food production. We kept activities going for as long as they were helpful in furthering discussion and creating connections between course concepts. Running the course on a natural sense of time, rather than the rigid structure I was used to, succeeded in inspiring students, at least in this instance, to think critically and understand how their experiences might lead them to make specific life choices.

The students in The Sustainability Semester were a self-selected audience of environmentalists who needed to understand communication and social persuasion to participate effectively in environmental deliberation. For this reason, I altered many of the activities in my course to include more practical components. Students wrote press releases and created storyboards for related news segments. They used their public education projects to complete work with their off-campus partnerships—writing white papers on mobile slaughter units, drafting resolutions for the creation of a local food policy advisory board, circulating petitions for divestment, and creating public art displays to encourage the rethinking of energy consumption. Our classes focused on audience analysis, message targeting, and overcoming communication obstacles. We also spent considerable time discussing how important it was to produce creative communication that breaks through society's message overloads. In other words, while my environmental communication course on-campus emphasized the environmental component of the course's title, the off-campus version emphasized the communication component.

Overall, the course evaluations demonstrate this approach was successful. One student wrote the course has "given me a greater understanding of environmental issues as well as how to communicate this understanding with other people." Another expressed, "I gained more perspective—overall understanding—more so than concrete information. What I think about in terms of communication is different now than before." None of these students came with communication studies or communications backgrounds, so conversations about successful communication and assignments that helped them to produce public environmental communication had a greater impact than similar discussions and activities had in the on-campus classroom.

The practical turn I took with the course allowed me to overcome two constraints I faced from this particular audience. First, many of these students were environmental studies and science majors who were not convinced the humanities had a considerable role to play in addressing environmental issues. However, as they saw how theories and practices of communication could be useful in bridging divides they had encountered in sharing their own environmental perspectives, the importance of the humanities and arts became clearer. In addition, these students were inundated with messages of sustainability every hour of their day. "Green fatigue" (Wilson 2007) was a risk, so I had to make sure that the course had enough overlap with their other courses so they could make connections, but not so much that it would be repetitive and useless. By emphasizing the communication component of the course and incorporating a synthesis discussion into each class session, I was able to (a) teach environmental communication concepts that added to their knowledge of sustainability, while (b) reinforcing concepts they encountered in other classes (i.e., anthropocentrism, nature–culture divide, sense of place, local–global, and environmental justice), and (c) modeling participation in productive sustainability discourse.

Teaching in this program allowed me to fulfill the educational mission Schwarze (2007, 97) outlined of "[e]nacting a problem-posing, problem-solving mode of inquiry," which produces deliberative pedagogy that moves students away from "existing communication practices to the invention of alternatives" and encourages the evaluation of these new forms of communication, guiding students in the practice of judgment necessary to move the world toward sustainability. Because the students were immersed in discussions of sustainability from a variety of perspectives, the program gave them opportunities to hash out the debates they were identifying between scholars they were reading. It also gave them space to form and practice holding their own positions in debates over sustainable practices and ethical environmentalism.

One of the greatest successes of the course was not something I saw in the course evaluations. Students confidently argued their positions, identified disagreements they had with me about particular philosophical debates, and raised challenges to my life choices, without fear they would be penalized for doing so. They became individuals capable of and willing to participate in productive deliberation over environmental policy and practices. Watching students grow into this identity helped me understand the importance of altering my pedagogical communication to meet the demands of the rhetorical situations of environmental communication classrooms.

Adapting environmental communication pedagogy

Collectively, the classroom experiences I have detailed here illustrate the importance of evaluating and adapting the environmental communication in environmental communication courses. The course's framing, structure, assignments, and content communicate to students the course's purpose and the roles the instructor expects students to assume both in and outside of the course. Each of these should be adapted to the exigences of the course and the needs and abilities of the students, the audience. I'd like to suggest some guidelines that may help instructors adapt their own pedagogical environmental communication:

- *Administer early course surveys:* A survey administered early in the course asks students to identify important environmental issues, explain global climate change, and note their interests and why they are taking the course. I use this information to tailor my lectures and activities to fill gaps in knowledge and make connections to students' interests. Environmental communication brings in a cross section of students who might not otherwise take a communication course or an environmental course. Determining which element of the course—environment or communication—needs to be emphasized and using the other element to maintain relevancy is essential in tailoring the course to the students' needs.

- *Use case studies:* Case studies can provide a course thread that connects otherwise abstract history, concepts, and theories. In addition, case study assignments allow students to see themselves as experts and gain confidence in producing environmental communication. These types of assignment allow students to use facts, figures, and other data, while connecting emotionally to environmental issues. Students remain engaged throughout the semester and see why they should continue to care about the environment.
- *Have students create public work:* Not only do students invest more when work will be seen by a public audience, but public projects mentor students in how to participate in public environmental discourse. Novice students recognize their ability to participate in public debate, and existing activists learn important lessons about audience consideration, message crafting, ethical argumentation, and utilizing creative mediums. All students have an instructor there to help craft the message before it goes public, and students have people willing to help shoulder the risk (namely the instructor) of people responding negatively. Public projects let students take steps toward performing environmental citizenship and success in those efforts can inspire them to continue enacting such a role.
- *Revisit the analysis of the rhetorical situation:* Throughout the semester make sure to ask questions: Have the students changed? Has what they need from the course changed? Have new constraints arisen? Vatz's (1973) critique of Bitzer's (1968) model provides an important lesson for communicators who practice environmental communication pedagogy. Meanings of rhetorical situations in the classroom are co-created by students and instructor. As early communication in the classroom changes the audience (and perhaps the rhetorical situation), checking in with the class to determine if the situation has taken on new meanings or has new communicative needs is perhaps the most important step environmental communication educators can do in improving their environmental communication.

Conclusion

Environmental communication educators inform, persuade, and encourage their students. Depending on the day, they might use informative, deliberative, *and* epideictic (ceremonial) rhetoric to meet the purposes of the course and needs of their students. Viewing this pedagogy as environmental communication suggests that perhaps environmental communication educators are doing a bit more than "enhance[ing] the ability of society to respond appropriately to environmental signals" (Cox 2007, 15). Environmental communication pedagogy trains students to be the responders. The environmental communication in these classrooms provides students with the information, practical skills, and encouragement they need to be citizens of and for the environment.

Andrew Dobson (2003, 176) argued that producing ecological citizens requires bringing together two strands of education: "education for citizenship, on the one hand, and environmental education (as it is normally called) on the other." Environmental communication courses pair the citizenship training of civic engagement and critical thought often present in communication studies courses with environmental education, and are thus uniquely situated to encourage the practices of ecological citizenship by students. When considering the role environmental communication courses play in training citizens, however, determining what type of roles students are currently prepared for and what they need to move toward adopting ecological citizenship is essential.

Our classrooms can be places to train active citizens. However, training activists is enhanced when the audience of students is ready to be mentored into that position. Educators committed to the ethical principles of the environmental communication discipline have to recognize this will not always be the case. Courses must be adapted to fit the needs of students and help move students closer to being willing participants in public environmental discussions.

In the tenth anniversary edition of *Earth in Mind*, David Orr (2004, xiii) expressed doubt in the ability of our current education system to "equip [students] to deal with the consequences of what we are leaving behind." Analyzing the rhetorical situation to improve the environmental communication of environmental communication classrooms may be a way to overcome some of the doubt Orr expressed about academy silos and a system entrenched in a "destructive and self-indulgent" society (ibid., xii). Education, he contended, must help students to recognize the problems of our current system and find a new way to live with the land. He explained, "We never have a choice to opt out of the land community. Our choice, rather, is whether to be a good member or a destructive one" (ibid., 209–10). Environmental communication encourages individuals to choose to be good members who use communication to persuade others to do the same.

Shaping our pedagogical communication to fit the demands of the rhetorical situation and talking with our students about how this approach influences the class throughout the duration of the course models the approach educators ask ethical and productive environmental communication practitioners to adopt. It involves listening, offering perspectives, altering approaches, and working together to create meaning, knowledge, and creative solutions to pressing problems. Educators ask their students to take active roles in environmental communication, so they must model for them how to do that. As educators use environmental communication to move students toward ecological citizenship, they must recognize themselves as the communication practitioners they are and meet the demands of the rhetorical situations they encounter.

Note

1 The author thanks her Environmental Communication students for their advice and feedback.

References

Bitzer, L. F. (1968) "The rhetorical situation," *Philosophy & Rhetoric*, 1(1): 1–14.

Carson, R. (2002 [1962]) *Silent spring*, Houghton Mifflin, Boston.

Cox, R. (2007) "Nature's 'crisis disciplines': Does environmental communication have an ethical duty?," *Environmental Communication*, 1(1): 5–20.

Dickinson, E. (2011) "Displaced in nature: The cultural production of (non-)place in place-based forest conservation pedagogy," *Environmental Communication*, 5(3): 300–19.

Dobson, A. (2003) *Citizenship and the environment*, Oxford University Press, Oxford.

Killingsworth, M. J. (2007) "A phenomenological perspective on ethical duty in environmental communication," *Environmental Communication*, 1(1): 58–63.

Leopold, A. (1949) *A Sand Country almanac, and Sketches here and there*, Oxford University Press, New York.

Masri, R. (2007) "Teaching amid despair: Global warming and Israeli wars on Lebanon," *Environmental Communication*, 1(2): 236–42.

Orr, D. W. (2004) *Earth in mind: On education, environment, and the human prospect* (10th edn.), Island Press, Washington, DC.

Schwarze, S. (2007) "Environmental communication as a discipline of crisis," *Environmental Communication*, 1(1): 87–98.

Sprain, L. and Timpson, W. M. (2012) "Pedagogy for sustainability science: Case-based approaches for interdisciplinary instruction," *Environmental Communication*, 6(4): 532–50.

Thoreau, H. D. (1908) *Walden, or, Life in the woods*, J. M. Dent, London.

Vatz, R. E. (1973) "The myth of the rhetorical situation," *Philosophy & Rhetoric*, 6(3): 154–61.

Wilson, H. (2007) "Have you got green fatigue?," *Independent*, 20 September [online]. Available at: www.independent.co.uk/environment/green-living/have-you-got-green-fatigue-402971.html (accessed March 21, 2016).

3 Environmental communication pedagogy

A survey of the field

Joy Hamilton and Mark Pedelty

Upon applying to establish *Environmental Communication* as a permanent course offering at the University of Minnesota, we received the following feedback: "The committee asked for a more accessible description of environmental communication … what is it, exactly?" (N. Hanks 2013, personal communication). This was, and remains, a good question. We know what environmental communication means to us, how we define it in our courses and research. After all, we have taught (or taken) the course for several years as a "special topics" course and feel we have a solid sense of what environmental communication is as a field of study and instruction.

However, do we, as a field, know what "environmental communication" is? Via common texts, journal literature, and discussion with fellow instructors, we gain a sense of how colleagues define the field. Some may reference Tema Milstein's (2009, 344) definition in the *Encyclopedia of Communication Theory*, which highlights the interdisciplinary tenet and ever-pressing concerns for "human-caused environmental crises." However, scholars and practitioners have their own unique definitions, as we came to learn through interviewing leading scholars in the National Communication Association's Environmental Communication Division for a divisional podcast (ECD 2015). For instance, Samantha Senda-Cook simply states that, "[E]nvironmental communication is how communication about the environment shapes our perceptions of the natural world and our relationships with it." Ross Singer makes a point to place emphasis on the nonhuman world in his definition. All are excellent definitions. Yet, few of us know what other environmental communication instructors *teach* in their courses and how they actualize the discipline in the classroom.

This chapter presents an overview of undergraduate environmental communication curricula based on a sampling of syllabi from US campuses. Our "map" of environmental communication curricula is intended to aid instructors in evaluating and advancing their courses. Environmental communication is a relatively new, but rapidly expanding field. With a topic as broad as the "environment," these syllabi represent a wide range of pedagogical approaches and foci. We hope this assessment will help educators identify gaps in the collective curriculum, indicating areas for future development. Additionally, we hope to create a bridge between research and pedagogy in order to advocate for a symbiotic relationship between the inextricably linked practices of teaching and research.

Similar studies have been conducted in other areas of communication and environmental studies pedagogy but not in environmental communication. For example, Morreale, Hugenberg, and Worley (2006) surveyed "the basic communication course" to determine which subtopics receive the most attention. Similarly, Kennedy and Ho (2015, 205) published a discursive analysis of twenty-two introductory environmental studies and sciences undergraduate syllabi, suggesting that instructors teach "students about not only today's environmental challenges, but also the comparative strengths and limitations of the possible pathways towards addressing them." Syllabi are the driving force behind developing students that are equipped to "critically evaluate and reflexively adopt environmental perspectives" (ibid., 202). In other words, introductory environmental studies and sciences syllabi should capture the diversity of thought in the field more accurately and ensure that students are taught the skills necessary to challenge various environmental assumptions. Kennedy and Ho's study parallels our project concerning environmental communication syllabi, specifically in their scope and objective to encourage instructors to continually re-evaluate their courses to develop critical skills. However, the aforementioned studies on pedagogy fail to go beyond classroom curricula to research; that is, their study's scope analyzes the strengths and weaknesses of collective syllabi but fails to draw a correlative to research within the same discipline; therefore, we compare curricula to research to understand syllabi in comparative perspective. That bridge is assembled through a basic comparison of topical coverage in the classroom and *Environmental Communication* journal articles.

We all wear multiple hats as researchers, practitioners, and instructors. These multiple roles prompted us to investigate the relationship between research and pedagogy. Therefore, our initial research question was simple: Which topics are covered in environmental communication curricula and research and how do the two relate? Through the process of deep reading and open coding of gathered syllabi, we identified several trends that highlight both common and widely divergent learning objectives, assignments, and pedagogical strategies. We highlight these findings with the goal of helping new instructors plan their courses and more seasoned teachers know where their environmental communication courses fit relative to the larger, collective curriculum in the field.

Method

We began by collecting syllabi through the International Environmental Communication Association (IECA) database or via listservs (electronic mailing lists) and direct email requests to IECA members and the National Communication Association's Environmental Communication Division (ECD). As the supply of graduate syllabi was limited, the study focused on undergraduate syllabi, and we chose to restrict the instructional years to 2010–15. In the end, we obtained twenty-seven syllabi.

After deliberating on how to best capture the objectives and methodological orientations, we concluded that assigned readings gave us privileged insight. Assigned readings were chosen for several reasons. First, syllabi vary radically in length and style, skewing potential comparisons of topical breadth across courses. For example, if one course emphasizing the environmental arts has a skeletal syllabus and one emphasizing science has extensive description of topics, the resulting coding outcomes would be an inaccurate representation of foci. The result would not necessarily mean that in combination more science than art is being covered by such courses in the aggregate. Coding the syllabi as if they were equal in length and format would lead to that mistaken conclusion. One of several side lessons of this study is that syllabi range vary so greatly in length and content as to make them barely recognizable as belonging to the same "syllabus genre." Conversely, all syllabi listed assigned readings, which are a good indication of topics covered in a given course and, as opposed to syllabi text, are amenable to comparative coding procedures. Therefore, we compiled all required academic readings, book chapters, and textbooks from the twenty-seven syllabi into one large sample. That sample most effectively reflects topical emphases in the collective curriculum.

Coding procedure

Two initial coding tasks emerged:

- all assigned course readings; and
- journal articles from the journal *Environmental Communication.*

We analyzed both according to ten pre-determined categories. Then, for additional insight into topical coverage for the twenty-seven syllabi, we used word-count software to identify subtopics instructors commonly incorporate. Finally, we used open coding to highlight pedagogical trends across the syllabi in relation to common assignments and course objectives. These findings represent what we found to be common trends and interesting approaches to teaching the course.

Because computer-assisted coding would not provide the connotative indications and nuance necessary for this study, the researchers coded the data themselves. Hamilton coded the data to determine the categories into which each reading fitted (e.g., a given article or book could fit several categories). The following pre-determined categories were used for the initial coding procedure. These categories were derived from chapter-level topics in Cox and Pezzullo's (2015) *Environmental Communication and the Public Sphere* textbook:

- Ecocritical Humanities
- Science
- Campaigns
- Arts and Popular Culture

- Journalism
- Advertising
- Environmental Justice
- Public Participation
- Professional Practice
- Other

To comparatively assess the results and have a yardstick to measure topical breadth and emphases of environmental communication pedagogy, Hamilton performed the same coding procedure on all 160 articles published in the *Environmental Communication* journal from 2010–15. This provided a comparative sense of the relationship between research and teaching—a relationship we find vital as researchers *and* educators (Figure 3.1).

Curricular subtopics and objectives

Naturally, any given coding scheme provides only a partial understanding of topical breadth and relative coverage. Our category scheme does not get at "environmental subjects," per se, such as "climate change" or "invasive species," "food," etc., that were the foci of several journal articles. To remedy this, we completed one final procedure using Linguistic Inquiry (LIWC 2015) and word count software (Friedman 2004) to gain a sense of how the collective curriculum and research might prioritize certain subjects and deemphasize others.

To do this, Pedelty collected all terms used more than ten times from the syllabi, minus typical parts of speech such as prepositions. These words were then categorized and grouped into like categories, while recognizing that this is just one of multiple potential taxonomies (Figure 3.2). We then looked at

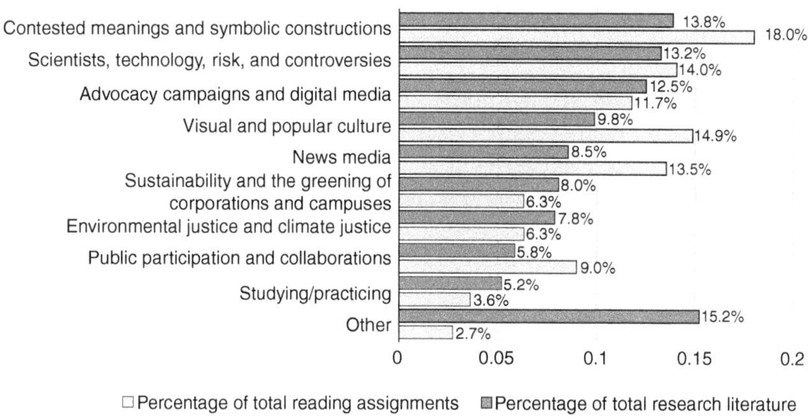

Figure 3.1 Proportional coverage of subtopics in environmental communication reading assignments and journal literature.

CATEGORY	WORD COUNT
SCIENCE climate (192), science (133), nature (113), natural (68),sustainability (63), green (61), energy (31), sustainable (23), warming (23), ecological (21), ecology (18), carbon (17), conservation (17), scientists (17), data (14), animals (11), land (11), sciences (11), wildlife (10)	854
SOCIETY public (214), social (107), community (42), ethics (22), history (21), society (20), ethical (17), behaviors (16), cognitive (10)	469
POLITICS justice (55), policies (45), advocacy (43), campaign (43), politics (43), risk (43), campaigns (42), political (27), movement (19), movements (18), activism (12)	390
RHETORIC rhetoric (46), messages (42), message (35), stories (34), audience (30), discourses (22), rhetorical (22), audiences (21), discourse (17), visual (20), interpretive (16), framing (14), story (14), messaging (12), frames (10)	355
CULTURE culture (72), human (39), popular (35), cultural (30), creative (25), communities (18), beliefs (13), ecocultural (12), gender (11), race (11), symbolic (10)	276
ECONOMICS marketing (34), client (31), professional (26), advertising (16), consumption (15), greenwashing (14), business (13)	149
NEWS news (73), journalism (25), reporting (17), newspaper (12)	127

Figure 3.2 Word frequencies and mega categories in environmental communication syllabi.

Source: author.

samples of word pairings to confirm that they are used as categorized (e.g., "communities" refers to human social groups rather than plant or animal communities). Pedelty removed terms that could not be disambiguated between categories (e.g., critical, critically, food, health).

While required readings and subtopics showcase methodological orientations and subtopics, we also wanted to explore other aspects of courses. For instance, what is the most popular type of assignment that instructors use? What are the most common learning objectives? To do this, we used emergent coding and highlighted several trends that we find interesting and relevant to both the field at large and our own interests as environmental communication instructors. Permitting pedagogical trends to emerge through repeated and close readings of the syllabi allowed for insights that were not pre-determined by the coder, thus opening up the possibility of unforeseeable outcomes.

Results

Topical coverage between reading assignments and *Environmental Communication* journal literature was relatively similar, as demonstrated by Figure 3.1. None of the topics surpassed a 20 percent coverage rate within their coding category (e.g., assigned readings or EC articles) (Figure 3.1). The coverage rate signifies the degree of coverage within the assigned classification (e.g., assigned readings or EC articles). "Ecocritical Humanities" was the top category, for both the reading assignments and the research literature. For required readings, "Sciences" was the second most common category, whereas for research literature, "Arts & Popular Culture" maintained the second-most common category. "Arts & Popular Culture" is the fourth most common category for required readings, suggesting that there is less emphasis on "Arts & Popular Culture" in the classroom than in research.

It is also worth noting that more assigned readings were coded as "other" (15.6 percent), whereas it is the least common for journal literature (2.7 percent). These readings were placed in this category because they did not primarily, or explicitly, engage with the communication discipline. The "other" readings for syllabi were diverse and ranged from classics such as Rachael Carson's *Silent Spring* to readings about storytelling, education, or psychology. Since several of these readings dealt with environmental topics such as DDT, food, environmental ethics, or climate change, we decided to find out how many syllabi explicitly stated that a course objective was increasing environmental literacy. The syllabi showed that approximately one-third (9 out of 27) of syllabi fell into this category.

To further explore categorical coverage in courses, we compiled word frequencies from all twenty-seven syllabi. Based on the frequencies of all words used in the syllabi, it would appear that climate change is the specific subject covered most often, with subjects like "animals" and "toxins," among others, mentioned far less often (Figure 3.2). Of particular note are the bottom ten mentions, where "gender" and "race" reside. These "bottom mentions" indicate peripheral topics instructors might consider incorporating in their course content.

Discussion

Before we began our coding procedure, our guiding hunch was that curricula would be less oriented toward science and policy than the research literature. That was not the case. In fact, it appears that environmental communication courses are surprisingly similar to the research literature in representing a wide range of the environmental sciences and humanities. In other words, it is a diverse, interdisciplinary, and multidisciplinary field, both in terms of teaching and research. We maintain that the field should remain open and diverse rather than become overly standardized and limited.

Figure 3.1 is a map of environmental communication terrain. It demonstrates that a relatively wide range of topics is covered in the aggregate curriculum and that curricular coverage maps onto the research literature fairly well. Of course, individual readings are skewed in various directions. For example, Cox and

Pezzullo's (2015) textbook has a bit more of a science and policy emphasis than the field as a whole. A simple search for meta-disciplinary references, using the search function in CourseSmart with an electronic version of the text, indicates that emphasis. The term "science" is used 144 times, a common theme throughout, as are "law" (122) and "policy" (108), while "art" is only used 36 times and "humanities" (4) and "social science" (1) are barely mentioned, at least in that more explicit form. While word frequency is a fairly crude cipher for understanding thematic emphases in a text, the large discrepancy might in this case say something about the orientation of the field's most common textbook and, perhaps, indicates the need to supplement it significantly when covering issues more directly involving environmental humanities and social sciences.

The word frequencies indicate that pedagogically speaking there might be less ecocritical and popular and/or visual material in the classroom than one might expect given relative coverage in research. The addition of a chapter dedicated solely to "Visual and Popular Culture" in the latest edition of *Environmental Communication in the Public Sphere* (Cox and Pezzullo, 2015) is another indication that the field is just starting to focus on those important subjects in earnest. Figure 3.2 supports this conclusion, given that the category "Culture" is invoked far less than "Science."

None of the required readings spanned across all or most syllabi. In other words, there is no consensual "must-have" text, although Cox and Pezzullo's (2015) book was used in nearly half of all courses. The second most frequently assigned textbook was Julia B. Corbett's (2006) *Communicating Nature: How We Create and Understand Environmental Messages*. Furthermore, the 15.6 percent of required readings in the "other" category might indicate that teaching is broader and less tightly defined than research as represented in the *Environmental Communication* journal literature. Several of the readings in this category suggest an attempt to increase environmental literacy of undergraduate students, with works from authors such as Michael Pollan and Bill McKibben. Readings in psychology, education, professional communication, storytelling, communication theory, and policy were also among common "other" readings. Additionally, some instructors assigned news stories from outlets such as *The New York Times* or websites such as 350.org. These types of reading did not qualify for coding (the study was limited to guides, academic readings, and books); they point, however, to a pedagogical strategy of assigning timely articles and campaigns that increase student environmental news literacy or objects for news media analysis assignments or discussions.

Emergent trends

With the goal in mind of helping instructors mold their syllabi, there are several other findings worth mentioning. Several courses are campaign-oriented: 51 percent of all syllabi either frame the entire course via campaigns, or have a major assignment that asks students to create their own campaign pertaining to their own topic of interest or related to a local environmental organization. This ranges from interpretive talks to documentaries or public education projects.

Some 33 percent of syllabi explicitly state the objective of increasing environmental literacy. In our experience, students in the Communication Studies Department at the University of Minnesota come into the classroom with varying degrees of environmental literacy. We consider that an essential element of our course, but it is a significant challenge.

Several assignments indicate that instructors are adopting a place-based pedagogical approach. Assignments such as regional documentaries and partnering with local environmental organization "clients" suggest as much. At the University of New Mexico, for instance, instructors design curricula that require students to directly engage with their immediate environment in Albuquerque. Other courses take students on field trips to nearby national parks or forests.

Place and identity emerges as a theme through assignments pertaining to autobiographies: twelve of the twenty-seven syllabi contained assignments that ask students to critically interpret their immediate surroundings and their own perceptions of "the environment" through personal narratives. Closely related to this premise were courses that incorporated field components. The "field" in this instance being any course component *outside* of the classroom. Roughly 30 percent of courses take group trips or assign student projects that involve hiking, listening and observing, tours, service-learning projects, or other "field" activities.

A field in the making

There are certainly many other ways to analyze the curriculum. For our initial question of topical coverage between courses and research, we concluded that required readings would provide the clearest insight into the curriculum as a whole. Of course, the method is limited, but we felt that given the newness of the field the first order of business was to answer the question "what are people doing?" at a basic level.

Identifying how the collective curriculum covers a given issue does not provide any indication of how individual instructors actually deal with that topic, nor is this survey of the curriculum intended to do so. Instead, we are attempting to provide an overview of environmental communication pedagogy as it currently stands, at this nascent stage of curricular development. At this juncture, the question "what is environmental communication" is still being worked out in the classroom, conference, and field site. These data indicate that it is a field very much in the early stages of formation. The diverse range of readings suggests that instructors are defining the field in unique and diverse ways. Most instructors are still determining largely for themselves what environmental communication is, and we argue that, at this incipient stage of the subfield, that is how it should be.

Clearly, as a field, environmental communication is transdisciplinary, but that is not always the case for individual courses. Some syllabi indicate a more disciplinary or even sub-disciplinary approach, with a few focusing on social science and others almost entirely in the rhetorical–interpretive tradition. Other courses, including those at the University of New Mexico, emphasize ecocultural and intercultural frameworks. We encourage instructors to peruse the IECA syllabi database to get a sense of the wider range of possibilities, as we did in preparation for writing this chapter. We also urge instructors reading this chapter to consider sharing their syllabi so we can all learn from reading about their approach. Finally, it would be insightful to expand the study outside U.S. borders as well as into the graduate classroom.

Conclusion

To summarize our findings, the environmental communication curriculum covers a wide range of subtopics and does so in a way that remains fairly consistent with the field as a whole. However, there is room for more expansion and equal distribution, as indicated by our word count assessment of syllabi. Assigned readings indicate that in the aggregate, at least, instructors are covering roughly the same subtopics referenced in the research literature, although the research literature seems to emphasize interpretive and cultural approaches to a greater degree than do most instructors' courses.

Our study highlights the breadth and diversity of the field. However, given limitations of time and space, this is more of a glimpse into an emerging field than a comprehensive overview of a solidified discipline. We undertook this research in part to answer a very idiosyncratic question: How does our course relate to others'? We suspect that both new and seasoned instructors might similarly be curious to find out what topics their colleagues cover and what approaches they take. We conclude that there remains significant room for experimentation and that a wide range of topics fit under the broad umbrella of environmental communication.

References

Corbett, J. B. (2006) *Communicating nature: How we create and understand environmental messages*, Island Press, Washington, DC.

Cox, R. and Pezzullo, P. (2015) *Environmental communication and the public sphere* (4th edn.), Sage Publications, Thousand Oaks.

Environmental Communication Division (2015) *What is Environmental Communication?* Available at: www.envirocomm.org/podcasts/ (accessed October 1, 2015).

Friedman, S. M. (2004) "Wordcounter." Available at: www.wordcounter.com/ (accessed October 20, 2016).

Kennedy, E. B. and Ho, J. (2015) "Discursive diversity in introductory environmental studies," *Journal of Environmental Studies and Sciences*, 5(2): 200–6.

LIWC (2015) "Pennebaker Conglomerates, Inc." Available at: http://liwc.wpengine.com/ (accessed October 20, 2016).

Milstein, T. (2009) "Environmental communication theories," in Littlejohn, S. W. and Foss, K. A., eds., *Encyclopedia of Communication Theory*, Sage Publications, Thousand Oaks, pp. 344–9.

Morreale, S., Hugenberg, L. and Worley, D. (2006) "The basic communication course at US colleges and universities in the 21st century: Study VII," *Communication Education*, 55(4): 415–37.

4 Breathing life into learning

Ecocultural pedagogy and the inside-out classroom

Tema Milstein, Maryam Alhinai, José Castro, Stephen Griego, Jeff Hoffmann, Melissa M. Parks, Maggie Siebert, and Mariko Thomas

As teachers in the relatively new, but increasingly in-demand field of environmental communication, we often find ourselves introducing first-time courses to our departments, programs, and organizations. This chapter argues that, as we design courses, we also have the related opportunity and connected imperative to turn conventional learning inside-out. Indeed, this book as a whole provides a crucial moment to reflect on ways many environmental communication teachers are doing just that and a moment to further refine these practices. In this chapter, we elucidate an *inside-out classroom* model (Milstein 2015) for teaching environmental communication, as well as environmental studies and the many related fields of learning. The model takes up Cox's (2007) ethical duty of environmental communication within the realm of pedagogy, teaching learners about our current global anthropogenic ecological systemic crisis from a communication standpoint while empowering them to be ecocultural[1] change agents. The goal with such an approach is to create transformative learning spaces in which learners' inner concerns and passions find vital connection with their understandings of, and practices within, the wider biosphere.

The chapter is written to mirror the pedagogical approach itself, as an exchange among the professor (Tema) and graduate students (all of whom have taken her classes and have taught their own university courses). Student voices burst organically throughout, nourishing and flourishing through interaction with one another and the instructor. As in a course, in this chapter these bursts can be read as rhizomatic, or as sprouts, buds, or blossoms that are the continued life of, but extend beyond, a teacher's original hopeful seeds of intention and design. In the following sections, we:

- illustrate an ecocultural orientation to pedagogy, which takes culture and its contexts as a starting point for understanding human ecological relations;
- describe the inside-out classroom model, its transformative goals, and specific techniques to achieve those goals; and
- discuss inside-out pedagogical practices of nurturing ecocultural change agents in our students and in ourselves.

An ecocultural orientation to environmental communication

Within the broad field of environmental communication (as well as environmental studies as a whole), an ecocultural approach uses culture as its primary lens in understanding human relations with/in/as nature. Culture is understood as a vivid window into human perceptions, embodiments, and practices of ecological relations, as well as a constitutive force of these relations. An ecocultural focus helps illuminate embedded cultural common sense, unearth forgotten knowledges, and open up material and symbolic spaces of possibility. Central themes include but are not limited to interpreting core ecocultural premises within systems of meaning (Carbaugh 2007a; Milstein 2016), as well as providing spaces of interruption of dominant ways of knowing, such as foregrounding the more-than-human world as co-communicant (Abram 1997; Rogers 1998; Carbaugh 2007b; Milstein 2012b) in constant processes of internatural communication (Plec 2012).

MARYAM: Learning with an ecocultural lens cultivated in me moments of disruption by provoking me to reconsider, reinterpret, and at many times reject prior assumptions and beliefs. I now ponder the questions: Of what stories do I find myself a part? The anthropocentric story? The conservation story? The ecological story? Outside, I now stop and reflect, instead of just walking by. Nature is alive. My ears listen to things they never paid attention to. I smell more things. I see more details. I feel more holistic. I now recognize and problematize culturally constructed binaries of culture–nature and subject–object, as well as knowledges shaped by hegemonic ideologies, such as capitalism. I have realized humans are not alone on this earth, that we harmoniously live with communities of animals, birds, insects, plants—and these realizations reflect messages I grew up with in Oman in the Quran.

As with much environmental communication research, an ecocultural focus helps strengthen understandings of communication as not only reflective of environmental relations but also as helping to (re)produce beliefs and practices and (re)construct identities and behaviors. Culture is seen as infusing environmental relations and as related to more or less equitable systems of environmental governance, economics, education, etc. As such, an ecocultural orientation both recognizes and questions power by, for instance, problematizing and challenging extractive ecocultural perceptions and practices that have become a globalizing status quo. At the same time, an ecocultural approach introduces and provides space for multiple culturally informed ways of knowing, and reveals underlying intercultural tensions and ideological struggles. In centering culture, especially their own cultures, in conversation with scholarship, learners often experience and share important interconnected realizations and/or alterations in ways they envision their own ecological positions in the world.

JOSÉ: In Western colonized spaces, such as my home country of Ecuador, the resistant notion of interculturalidad as opposed to interculturality (Aman 2014) has fostered inter-epistemic communication (Mignolo 2007), which problematizes the position of humans as the source (center) of knowledge because it embraces the idea of nature as the source of knowledge. Therefore, nature can teach us. Within this interrogation, human–nature relationships are seen as a membrane of relationality, complementarity, integrality, and reciprocity. This helps us remember that we start from everywhere and are going in many directions at once.

The inside-out classroom model

The transdisciplinary greening of the university (Fassbinder et al. 2012) has challenged educators to meaningfully green their own pedagogy. The inside-out classroom model is one response and takes its name inspiration from the evocatively named "flipped classroom" (Tucker 2012), which precedes class time with online lectures and reading discussions to reserve meeting time for hands-on high concept engagement. The inside-out classroom similarly relies on deep in-person experiential learning. However, its core emphasis is on cracking open conventional pedagogy to let relevance flow inward and outward, breathing life into learning at a time when education is poised to strongly impact quickly shifting ecological realities in which we, learners and instructors, find ourselves as actors and potential change agents.

The inside-out model focuses on ways to respectfully open and connect learners' *inner* life experience—their queries, understandings, abilities, passions—to course concepts and frameworks, and to extend knowledge building and engagement *outside* classroom walls and institutional boundaries. The approach deconstructs separations between inside classroom and outside world, between university and place, humanity and nature, body and environment, self and other, etc. The model is nurtured by the radical roots of problem-posing pedagogy initiated by Freire (1970) as well as ecopedagogy's integral transformative learning, which emphasizes not only the importance of questioning and critiquing but also on empowering agents of change (hooks 1994; O'Sullivan 2002; Kahn 2008; Milstein 2012c).

Linking internal and external relevance, as well as transformative possibility, means seeking ways to dissolve classroom walls, to empower learners to co-create their education, and to stimulate inclusion of multiple perspectives and diverse voices. Implementing the inside-out model also demands combining pedagogical flexibility with rigorous preparation and organization so students feel reassured they can step beyond the conventional learning boundaries to which they are accustomed. When the classroom cracks opens, students begin to root learning within the fertile soil of their own life experience, and to integrate their education into personal and collective growth, and to branch out far beyond the confines of the university.

The inside-out classroom model attempts to put into pedagogical practice environmental communication's ethical duty (Cox 2007) by creating a powerful launching process to engage learners' inner concerns with wider biosphere issues and to collaboratively gain critical, creative, and practical tools. Inside-out objectives and strategies have transdisciplinary and transnational applicability and are responsive to specific types of knowledge that contemporary universities are engaging and producing.

JOSÉ: Greenwood and Levin (2005) identify three types of knowledge at play in today's universities: 1. "Tacit knowledge" connotes the hidden understandings that guide our actions without our ability to explicitly communicate what those understandings are; 2. "Knowing how" grounds knowledge in actions and, in so doing, anchors tacit knowledge; and 3. "Collective knowledge" is formed by the knowing together of three kinds of equally valid knowledges (Aristotle's taxonomy): episteme (theoretical knowledge), techne (pragmatic and context-dependent practical rationality), and phronesis (practical and context-dependent deliberation about values). My experience is the inside-out approach attempts to make explicit tacit knowledge by embracing "knowing how" via praxis and lived experience, and, in the process, generating collective knowledge.

JEFF: Embracing collective knowledge and making tacit knowledge explicit are also direct routes to disrupting the enclosure of the commons, both in the literal and metaphoric sense. Traditional classrooms tend to close off our bodies from interdependent relationships with other ecosystem elements and teach us to think and act in ways that partition the world and center ourselves, rather than seeing the inherent interrelatedness and interdependence of systems. In traditional classrooms, we often embody the logic necessary to support extractive and destructive ecological practices; we distinguish our own bodies from others, our own experiences from ecocultural relationships that make those experiences tacitly real, and our own minds from our interconnected bodily knowledge.

Turning the inside out

> "The form of action they adopt is to a large extent a function of how they perceive themselves in the world."
>
> (Freire 1970, 83)

Turning the inside out centers first-person cognitive, embodied, emotional, cultural, ecological, and (for some) spiritual reflection to assist learners in meaningfully locating themselves as interconnected with/in a more than human world. Strategies amplify learners' inner queries and lived experience and ground the learning of course concepts in ways relevant to their personal journeys as learners and human beings. Examples of effective and widely applicable specific inside-out pedagogy techniques include free writes, co-teaching, and open final project menus.

MARYAM: Tema discusses these strategies, but others, too, were important to me. For instance, a first day icebreaker that had us asking each other questions, such as whether we knew where our drinking water came from, was an eye opening experience. It was my first encounter ever with my ecological self. I realized how cut off I am, which was a realization mixed with guilt and responsibility. How come I drink water all the time but I never questioned where it comes from? The activity increased my understanding of the web of life and how we humans are interdependent with/in nature. It also encouraged me to question how our social conditioning reinforces separation and distracts us from the fact we are a part of a whole.

The strategy of an in-class guided free write is a simple, highly effective way to help learners tap into their inner questions and concerns as they relate to core concepts. Free writes can be used to begin the term to establish the foundation that student perceptions and concerns matter and exhibit how they connect to core class questions, as well as to end the term to provide learning reflection and closure. Free writes help turn perceptions inside-out, making ecocultural ideologies and personal orientations visible, providing solo and shared space for reflection, and offering an inner touchstone and site of transformation throughout class as we grapple with complex topics. Elbow (1989, 46, 61) argues free writing "invites a personal honesty even in academic writing" and helps one "break free from what feels like the heavy mud and clinging seaweed" that clog one's ability to say what one already knows.

During a first or second class meeting, using the prompt "When I say nature, I mean...," learners can write nonstop for a few minutes without correcting, erasing, pausing, or worrying about what they write. This activity evokes a stream of consciousness that overcomes inner editors and locates deeper meaning systems and questions. After three to five minutes, the instructor halts writing and asks learners to underline one word they wrote that most means "nature" to them. Together, the instructor then facilitates each learner saying her or his word, creating a river of words, one after the other, with the instructor writing these on the board. Follow-up discussion helps participants connect their own solo and collective stream of consciousness to theory and wider contexts: Is "nature" difficult to put into words and, if so, why? Is "nature" the same for everyone? What is *not* "nature"? What is the dominant meaning of "nature" in this culture? How might your free write change if prompted with different terms such as "When I say 'the environment' or 'ecosystem' or 'Mother Earth' or 'environmental resources'"?

After group reflection, space can be created for (re)envisioning: participants can come up with terms and associated meanings for "nature" they feel comfortable with for now and write them on the board. This simple exercise often leads to potent experiences of revealed, and at times overturned, cultural perspectives, with international, indigenous, minority, majority, traditional, and nontraditional students obtaining shared space to multiply standpoints informing the class. At term's end, when we do the same free write again and compare responses, the group can reflect on ways and reasons their own meanings of "nature" may have changed or stayed the same. In the process, learners have the opportunity to witness the fluidity and fixity of environmental communication in ourselves. Starting and ending with this exercise helps create space for inside-out awareness and learner-initiated problem-posing, providing moments in which participants can creatively encounter, and at times trouble, their fundamental ecocultural beliefs.

Just as teachers are always learners, learners are always teachers. Another strategy of inside-out pedagogy is student co-teaching, which foregrounds this dynamic and provides guided support to collaboratively explore class concepts and develop a sense of investment, confidence, and camaraderie. Co-teaching at the graduate level both values and fosters student pedagogic voice and ability. Instructors who teach in programs with no environmental courses can create a few open theme weeks in more general courses for students to co-teach on related topics they identify as dear to them—for instance, in past years in a graduate Critical Cultural Studies course, students have chosen to focus and co-teach these open theme weeks on ecofeminism, indigenous environmental meaning systems, environmental justice, etc. Students can closely work with the instructor in twos or threes preceding the week to collaboratively plan and teach a lesson. In graduate environmental communication courses, students can be responsible for co-planning (also with instructor feedback) an "Application Activity," in which they design and facilitate a half-hour experiential activity that emerges from their own interests and strengths as well as that week's readings to help bring readings to life. At the undergraduate level, students can have voice as teachers through approaches ranging from small but meaningful acts, such as including a regular time at the start of each class for their announcements of course-pertinent wider environmental events, to substantial roles, such as teaching a particular skill or area of knowledge in which they have expertise. Students at both levels experience a deeper level of immersive learning from co-teaching opportunities, and, working with them in preparation and delivery, the instructor is able to support them in developing as teachers themselves.

Finally, an open menu of final project assignments from which students can self-select and meet learning objectives allows learners to tailor projects to inner queries and strengths. The menu can include conventional research papers, creative environmental advocacy efforts, educational websites or video documentary projects, research-service-learning projects in the wider community, public ecoculture jams (Lasn 2000), etc. Such openness in assignments often challenges student expectations and pushes their comfort levels. To address this and support all learners, in-class peer workshops leading up to the projects allow students to propose ideas, get and give feedback, collaborate by choice, and fine-tune proposals, while the instructor gives ongoing feedback at different stages of development. When given the opportunity to choose their own form of engagement with course material, students tend to be energized about their projects and often produce high impact, rigorous work, including projects that have effects beyond the classroom walls.

Final projects can be further turned inside out by offering public forums for students to present their projects. Students consistently take up this opportunity, and collaboratively decide on presentation formats, which have ranged from panels to gallery-style presentations to performances. At times, students have presented in public settings, such as the student union building or an art gallery, at others they have invited the public into the classroom. The opportunity to engage the public helps students see how their environmental communication work might have value for and impact on the wider world.

In addition, students who choose ecoculture jams from the open final project menu enact these projects in and with the public and post recordings of their jams to Youtube, expanding their reach. This extension of learning turns their individual creative activism into a shared and publically engaging performance with the power to relocate the ecocultural status quo (Milstein and Pulos 2015). Culture and ecoculture jamming are methods both for interrupting dominant narratives and discourses and for engaging in creative and playful forms of re/envisioning. In the inside-out approach, students have done ecoculture jams in spaces ranging from city bus stops to shopping malls to campus lawns. Examples include:

- a student who brought a live turkey he had raised and intended to slaughter to the student union building the week before Thanksgiving to engage anyone who approached in conversation about industrial factory farming;
- a student who created a dress to which she attached her garbage as she accumulated it throughout the day to help herself and others encounter and embody the impact of their trash; and
- a group of students who challenged, and re/envisioned, fast food chains on campus by creating a prank corporate booth and posing as Chick-fil-A representatives handing out freshly blended veggie smoothies and surveying delighted passersby on whether they would support the chain in shifting to a healthy, local, organic, and vegetarian menu.

Turning the outside in

> "Embracing the inward path to attain outward connection..."
>
> (Castro et al. 2015)

Inside-out pedagogy also means grounding learning outside classroom walls and embedding learning in wider relevance. From a practical standpoint, this encourages getting outside to observe, participate in, and better understand one's ecocultural systems. Three overlapping core techniques—field studies (group and individual), research-service-learning (RSL), and simply moving learning groups outside—serve to root learning in local practice and place, centering first-person experience in outward engagement of course concepts.

Immersive field studies apply course concepts to everyday or extraordinary sites, with the more mundane sites helping students see the familiar anew and the more unusual taking them out of their cultural milieu and expanding their reach. Field studies can range in effort output from instructor-designed but self-led individual studies to multi-day instructor-led excursions.[2] Field studies center both community members and place as teachers, challenging the assumption that legitimate or valid knowledge solely resides in classrooms, textbooks, or professors. Field studies also expose learners to wider contexts and serve to bond and activate. Each individual field study assignment can include a course concept-based question guide and data collection tips. All can be followed by reflection papers and in-class discussion, in which learners are asked to use first-person voice to put holistic field observations (cognitive, emotional, sensory, embodied, internatural) in conversation with course concepts and readings. Reflections help the instructor assess ways students are able to interweave their learning within the wider world.

MARYAM: The embodied experience of fieldwork greatly influenced my learning of abstract concepts. It developed in me a fertile mind that is ready to learn, relearn, co-learn, and unlearn ecocultural concepts. Overall, the field study and reflection not only enhanced my academic growth but also my personal growth. I developed confidence speaking about nature, culture, and communication. At the same time, I came to realize how little the ecocultural toolbox of either my home culture or my host culture is equipped with ecocentric vocabulary.

Instructors can heighten the experience of "turning the outside in" by asking those to be visited how the class can be of service or give back during their field study visit—for instance, lending a hand in ways needed at that moment during a visit to a community site of environmental labor. In some classes, however, starting with the primary focus as RSL offers experiential engagement that brings both personal investment in the wider community and environment and break-through moments. In highlighting the value RSL provides, students discuss indelible experiences that transform their thinking; at times, they make these the basis for enduring commitments. Some continue to volunteer, take on internships, or write dissertations or theses with community sites after class, and some seek out other opportunities to learn while being of service. As with field studies, RSL experiences are preceded by course-concept-focused question guides and followed by reflection papers and discussions in which students grapple with course concepts as they are grounded in their experience, and discuss ways service-learning may have confirmed, challenged, or changed their ecocultural perceptions.

MAGGIE: Field studies and research-service learning take our notion that we only must look to our instructor for the answers/knowledge and turn it inside-out. When I teach our university's community gardens and communication course, a course I first took as a student with Tema, I have learned to preface my instruction by telling students on the first day I do not pretend to know everything. I am learning, too, right alongside them as we work on the campus food gardens and with six farm and food garden sites we visit throughout the city. I am there to act as a guide toward mutual discovery. After all, community gardening is a skill/process: what works one year may fail the next. All the sites do things completely differently. Some use permaculture techniques to maximize yield in minimal space creating a cyclical regenerative growing space, whereas others are large-scale commercial operations replete with aquaponic set-ups and cute baby pigs that eat the leftovers. Students pick up all sorts of new knowledge and information that a textbook or I would not think to incorporate. At each site we get our hands dirty assisting farmers: we weed, plant, dig, compost, mulch, sweat, learn, and grow together. While the inside-out classroom is about more than just taking our classes outside, when we literally take our classes outside we create a context to question the supposition imposed by classroom walls—that learning happens inside, not outside. When we take our education outside, students take in a deep breath of the world around them; they look around, feel the temperature, spread out on their bellies, and relax a little. It seems to automatically shift the learner into a space where they don't have to "think" so hard and, instead, let what is going on around them inspire fresh thoughts, like breaths of fresh air.

Indeed, grounding learning outside also can be as simple as moving a classroom activity or discussion out of the building or giving learners a way to do so on their own. For instance, instead of simply having students pair and share to warm up for a class discussion, having them exit the classroom to do the same while walking and talking for five minutes outside creates a re-energized and outwardly connected learning space. Many of us who teach environmental communication and related environmental studies tend to be the first in good weather with our classes outside sitting on campus lawns or under shade trees; the simplicity of this move outside, however, may conceal its potential transformative power in pedagogy and learning.

MARIKO: It's 2 p.m. on a Thursday, and I can see my students' eyes rolling back, the more energetic ones frantically trying to text underneath the table with their eyes darting from me, to the Power Point, to their phone, and back to me (glaring) at them, while others fight the exhausting glow of the fluorescent lights to focus. This is not an abnormal scene for any teacher, and most of us in this position can feel the manic distraction that happens in white-walled rooms with isolating rows of desks. "OK, this is ridiculous; everybody, up," I demand mid-lecture, and their eyes blink in sedated surprise. "We're going outside." We resettle underneath an immense sycamore, and the circle of bodies nestled on the ground instantly becomes both more alert and relaxed. I restart the lecture, and student questions and additions become engaged more directly with one another, hands digging in the grass as they ruminate on their talking points instead of idly pressing keys on a laptop. The ones who struggle with paying attention zone out briefly, but, instead of cell phones, they focus on the light streaming through leaves and the knobby stick rolling in their fingers. It's the kind of zone-out where they are still listening and present. I smile because sitting in a circle outside with them also calms me down and lets me tune in to what they are trying to say, instead of authoritatively reconstructing their discussion points toward answers I seek. My experience as a student with inside-out pedagogy has allowed me to think more creatively about my own pedagogical role and what a "classroom" is and to let go of any authoritative ego I may have thought was necessary to make up for perceptions about my youth or gender. I still remember my first environmental communication graduate class where I was delighted and shocked to find myself sitting in the very spot to which I later brought my students, percolating on theories and figuring out with the group where I stood on them. I now take this expanded understanding of mitigated instructor authority and of what a classroom is into my own teaching, acknowledging gratefully that we can create a learning space anywhere. A class is a co-constructed space dependent on the relationships present and, no matter the course, a little sunlight and grass underfoot can be the best classroom technology available.

MELISSA: Before graduate school, my home for more than two years while I was a Peace Corps volunteer was an isolated, rural, jungle village in Malawi, teeming with life, sound, dirt, blood, sweat, bugs, steaming rains, creatures that could eat entire structures nearly overnight. Returning to the United States, with its concrete and traffic, was hard enough. Returning to the classroom, with its fluorescent lighting, sterile floors, freezing air conditioning, and toxic clean scents, with its professional dress code and $400 textbooks … that, I thought, might be enough to absolve me of a soul. I tried hard to embrace it, to become a person who could thrive in that environment. I was in a new city in what had become an unfamiliar national "culture," and I threw myself into the effort of conforming. For about four months. Just as I broke, I stumbled into inside-out learning and an ecocultural approach, and I found a family of like-minded (and intelligent, humorous, hopeful, hardworking, imaginative, energetic, positive) human-animal-student-teachers and, as spring began to creep into our town, so my wilder priorities began to tiptoe back in. Yes, I need and want an education, but I refuse to do it inside my cardboard cubicle under artificial lighting for years on end, blinding myself before a computer screen. Like water, like food, like sleep, I need sunlight. I need wind, rain, grass, dirt, movement, and adventure. Give me a ditch to dig, a cave to explore, a road trip, a journal, and excited conversation with like-minded people around a fire. THIS is learning. THIS is the way to connect theory to life. THIS is what I was seeking from my graduate education, and I have discovered it. A reprieve from the fluorescents. A way to look beyond the concrete. Salvation from, and yet in, an academic world.

We hope it is clear by this point that the inside-out classroom means more than merely moving learning outside classroom walls, though this is a fertile start. The inside-out model aims for holistic and interconnected learning experiences, which center not only intellectual learning but also cultural, experiential, emotional, phenomenological, sensorial, embodied, spiritual, and seasonal learning. Every place in which one teaches offers its own ecocultural complexity and wonder, situations and locations in which learners can engage, thrive, and identify their distinctive roles as both participants and change agents.

Nurturing ecocultural change agents

Anyone who has been fortunate to teach an environmental communication or environmental studies course knows students often germinate passion and intention in these classes. Some advanced undergraduate students and many graduate students arrive in full bloom. Our job as teachers is to help them reach stages of fruition. This means helping them root in the fertile soil of their inner wisdom—and the potentially sacred space of a class term—and reminding them of their interconnected existence, and efficacy, within wider integral ecological and cultural systems.

A mix of high teacher expectations for one's students and one's self, combined with a shared hunger for conscious engagement in our current ecocultural situation, helps all feel free to expand from the role of active learner to the role of conscious actor. All participants can recognize the rare opportunity to "do" such courses provide. For example, one undergraduate in a recent course evaluation wrote, "We are doing really important things that other people on campus don't have the privilege or perhaps yet even the awareness to do." This potent sense of opportunity combined with empowerment can be nourished from the inside out by helping students deconstruct conventional educational models they have been part of and giving learners an agentive role in creating their learning experience and their experience as change agents in the wider world.

One technique for helping to nurture change agents from the inside out is to have students learn critical questioning by starting with their own immediate environment (the classroom) and then expand outward (the university campus, the city environs, the "wild"). For example, one can facilitate students through the critical thinking steps of describe, interpret, and evaluate via a deconstruction of the classroom as an ecocultural artifact. As they describe, students might note the right angles, blandly painted walls, desks positioned to face the teacher, the absence of any notable plant or nonhuman animal life, the lack of open windows, desks oriented to face away from views outside. As they interpret, they identify themes and patterns in ecocultural spatial communication and connect these to what is valued as educational and as not educational in conventional higher education. Student responses often speak to how today's conventional classroom is, in many ways, modeled after Foucault's (1977) concept of discipline, in which systems of authority limit and control bodies' movements and experiences of time and space. Discipline requires and reproduces hierarchical observation, normalized value systems (and value judgments), and the absence of interconnection with/in the more-than-human world. With the instructor's support, students then evaluate what their interpretations might mean for ecological relations on a wider scale and consider actions to remake learning spaces to be more ecocentric and less disciplining. In this way, they move to and through awareness building to creative re/envisioning.

JOSÉ: Rediscovery is another way of describing the influence of using inside-out pedagogy. Following the tacit inside-out tenets, I ask my students to unplug for two days from all technological devices. They are allowed to check e-mail only to keep up with school or work. In reflective essays students turn in, they use the word "zombies" to describe how they perceived others "attached to," "enslaved by," and "shielded by" their cell phones. I take one class session to debrief their experiences and present a selected cluster of words or phrases that were recurrent in their essays to open discussion. The result is amazing not only because of the self-reflexivity the exercise triggers in them, but also due to actions they decide to take about technology use, i.e., cell phones. Several think about quitting regularly, perhaps once a month, as a way of "detoxing"; others have shifted habits to self-regulate phone use; many have challenged family members to do the same in order to recover spaces "to have a decent conversation." In terms of ecocultural experience, the epiphanies are about the senses and reconnecting: "I noticed...;" "I just listened...;" or "It was the best meal ever..."

During the first meetings of a course, after deconstructing conventional learning spaces and experiences, one can start facilitating students in consciously co-creating their own class culture. To do so, one can ask learners who are gathered to remember and/or envision their ideal learning experience. Then, facilitate an open brainstorming to share and build a list of elements of this experience (e.g., respecting diverse ideas, laughter, seeing how my learning connects to real life, arranging our seating how we want each day, no one talks too much or over others). Then, the instructor can have the class vote on which aspects to include in the class culture (and they can vote on how votes count. Unanimous? More than 50 percent?). The process creates a transformative learning space in which all can mutually experience the process of co-creating a reality, and helps all participants comprehend the learning environment as a co-constructed space for which they are also responsible. In creating this space, the instructor also facilitates participants in coming up with ways to operationalize or perform the consciously created culture, practices that can influence everything from in-class experience to syllabus structure. The process not only invigorates student empowerment but also helps them critically reflect on conventional learning spaces and become deeply engaged and dedicated in their learning as they have an opportunity to mindfully practice their culture throughout the term. The class culture list can be posted on the class website and the instructor can facilitate check-ins during the term and at its end, re/minding and reviewing together whether and how the class has lived by its co-created culture and whether any adjustments can be made.

MELISSA: Our graduate environmental communication class culture list included such items as not qualifying or apologizing for our contributions or questions. No more "I might be wrong, but..." or "This might sound silly, but..." I've been in classrooms where teachers attempted to set this precedent, but, for the first time in my experience, the resolution was actively introduced, supported, and executed by the students. Since we had co-created our culture together, and collectively decided to follow this standard, it quickly became the accepted norm to speak with confidence (and to support and encourage this, as listeners). The shedding of protective, defensive clarifications and apologies was a monumental experience for me as a learner—I found that speaking as if I knew what I was talking about meant I actually did. I had stumbled into my own unique brilliance, and I began to value my own perspective. In the safety of this classroom culture and with the support of my peers, I could freely ask questions—clarifying concepts, challenging topics, and shaping my interpretations. To be in a space where I could unapologetically practice this vital skill was of immense value. The confidence I gained continues to trickle down and through every aspect of my life, drastically improving the quality not only of my learning and teaching, but also my everyday behaviors and overall self-image.

The class culture exercise also inevitably generates fresh pedagogical ideas as each new class is shaped by the new learner group. Many of these student-generated ideas work so well in turning pedagogy inside-out and in nurturing ecocultural change agents, they are incorporated into future classes. For instance, the idea of walking and talking outside as a substitute for the pairing and sharing in one's seat that often precedes a large group discussion was collectively created by an undergraduate class during this exercise, as was including a time at the start of each class for students to announce wider ecocultural events. Similarly, in co-creating their culture, a recent graduate class came up with the idea of "talk the walk."

JOSÉ: We decided when we formed our class culture to set a time aside at the start of each class to "talk the walk," to voluntarily share our hopes, struggles, frustrations, and achievements in ecocultural changes we have chosen to enact in our lives. For instance, after looking at the food industry and its environmental impact, some of us embarked on the journey of becoming vegetarian or vegan. This inside-out approach helped us "ventilate" challenges in class through open conversation. These experiences in many ways shifted our bodies, helped us question naturalized ways of living, and fed our growing environmental awareness and beliefs that we can enact change.

Inspiring change agents also means pulling learners out of the present moment to provide perspective that, in itself, can be empowering. On a scale broader than the classroom or our technology bubbles, an ecocultural orientation can provide such perspective by simply locating the human in ecological and geological time. One move is through visualization. For instance, classes can begin by locating the human place in Earth time and geological history. Evocative imagery, found easily online, includes a geological clock face or spiral that helps visualize the scant time humans have been around in the scheme of life of the planet. A real-time flip-counter found online shows human populations increasing along with deforestation, industrial production, military spending, extinctions, and Earth temperatures and the same-time decreasing of biodiversity, wildlife habitat, etc. After using these visualizations both to locate and decenter our species, we proceed to explore the notion of the Anthropocene, as well as the more critically informed notion of the Capitalocene (Moore 2015). In the process, learners develop their own understandings of their unique opportunities (and often their ethical duty) to be agents of regenerative ecoculture in our time.

MARIKO: Classrooms that focus on environmental communication have the tendency to bring up heated topics closely bound to ways students understand their identities, culture/s, and the universe, as well as their place nestled within it. Like any large shifts in life, the act of exposing students to ideas that directly challenge values they consciously or unconsciously have held close to their hearts throughout their lives can be a painful process and, if done insensitively, can make students feel attacked in their beliefs and powerless in their opinions. The inside-out approach to dispersing power among a class of learners and building interpersonal communities soothes this fear and defensiveness. Students know their learning community is supportive, accepting, and present. Additionally, it encourages them to go through the processes of shifting environmental ideologies with agency and support, without feeling like it was forced upon them.

Bursting forth

As we were completing this chapter, Tema received an email from an undergraduate student, reflecting on her learning experience after the end of term. The student's sentiments of growth illustrate the cultivating power all environmental communication and studies courses have in learners lives:

> You can't grow a tree overnight just as you can't communicate environmental protection in that same amount of time. In order to grow a tree, you must first plant the seed and then foster it with love and care in order for it to thrive. Communicating a sustainable lifestyle is extremely difficult in a country where living a life of excess is most people's goal. Teaching a course such as this one essentially makes you one of many "knowledge gardeners." The idea of sustainability and concepts of preservation were planted as seeds in our brain only weeks ago at the beginning of the term. The thought grows as we continue on in our lives and is spread to others through us. As the knowledge spreads and the seeds grow, there could be nearly 7.3 billion plants in your garden.

We close with such openings: with graduate students, soon to be faculty, reflecting on their overall engagement with turning learning inside-out, fleshing out this ecoculturally infused way of learning and teaching, and illustrating ways their learning and teaching is shaping the fruition of their own paths and the myriad interconnected lives around them.

JEFF: One of the most powerful effects of the inside-out classroom, specifically as it relates to environmental communication pedagogy, is the potential it provides for seeing our own bodies in terms of other aspects of ecosystems (or getting us as close as possible while still respecting the wholeness of other beings, elements, and forces of ecosystems). We so often describe other ecosystem elements as either symbiotic or parasitic, predator or victim, hunter or hunted, supportive or destructive, nurturing or foreboding, etc. I have begun to think of all the ways in which my own interrelated presence in ecosystems is always agentic on both sides of any of these binaries, and that, in processes of rediscovery of the "embodied commons," all of us can begin to make more ecocentric choices.

JOSÉ: The inside-out approach is an integral pedagogy. It is not only a toolbox but also a frame by which to think and feel. I have heard that an instructor's behavior is different in or out of the classroom. But is it? Maybe for other styles of pedagogy, but not for this one. Inside-out pedagogy tries to break down the human/nature dualism; with the same intensity, it also questions the in-class/out-of-class instructor's performance dualism. For instance, how can I talk about the pollution generated by plastic bottles and show up every class with one? To what extent can I address the environmental impact of overconsumption and food waste if every time I go out to dinner I ask for a Styrofoam to-go box and maybe later throw out the food? How conflictive is it to encourage students to recycle when in your neighborhood there is no recycle bin so you just put everything in with the trash? Inside-out pedagogy thus is an ethical pedagogy that exposes your inside to your students (and the world). In a Freirian philosophical sense, to be is to become. Hence, inside-out pedagogy not only offers pedagogical tools, but most importantly, a means to perform and practice inside-out. This entails acting against the academic schizophrenia of being one person in-class and another outside.

STEPHEN: I have had the opportunity to participate in many transformative activities in the inside-out classes I have taken as an undergraduate, a Master's student, and now a PhD student, and they have had an effect on me from the start. As a returning student, I rediscovered my passion for academia and teaching by taking these classes. Over time, the philosophies behind these methods have fundamentally transformed my worldview. The inside-out classroom provided a safe and supportive "place" conducive to reflection of thought and empowerment of action, transforming learning from a singular pursuit into a shared experience that embraces culture, community, and the natural world. This semester marks the end of my second year teaching in a university. My first was admittedly a struggle dominated by my self-imposed need to develop a teaching style. I also found myself constrained to conventional textbook instruction conforming to a transfer-of-knowledge model, yet I was constantly seeking to disrupt this model with instances of inside-out methodology. It is in these disruptive instances that I began to find the seeds of my own style, as I came to understand the inside-out classroom was most effective when rooted in one's particular inner strengths and queries and grown in conversation with one's students, as a holistic methodology rather than a singular strategy.

Notes

1 Compound terms, such as ecoculture, humanature, humanimal, among others, reflexively embed ecology and culture, human and nature, human and animal in mutual conversation as they are in life. These symbolic moves are turns away from binary constructs and notions of "the environment" and toward a lexicon of reciprocity and reunion (Milstein 2011, 2012a).

2 Field study examples from Tema's recent classes include a one-day instructor-led undergraduate field study that spans the city from ancient indigenous petroglyph cliffs on Albuquerque's west side, to environmental justice work with an organization in the low income and undocumented immigrant city center, to prairie dog colonies surviving high income development on the city's east side, to foothills of the mountains that halt urban growth. Self-led studies include "constructed" nature, such as the zoo or a city park; six weeks of self-chosen sustainable change such as not using a car or going vegan; and a solo experience in "wild" nature. A graduate example includes a two-day instructor-led field study of northern New Mexico that engages us in ecocultural places such as the holy dirt pilgrimage chapel of the Santuario de Chimayo, Taos First Nations Pueblo, the Earthship Biotecture community, a Nuevo Mexicano agricultural nonprofit, Ojo Caliente hot springs, and Bandelier National Monument.

References

Abram, D. (1997) *The spell of the sensuous: Perception and language in a more-than-human world*, Vintage Books, New York.

Aman, R. (2014) "Why interculturalidad is not interculturality," *Cultural Studies*, 29(2): 205–28.

Carbaugh, D. (2007a) "Cultural discourse analysis: Communication practices and intercultural encounters," *Journal of Intercultural Communication Research*, 36(3): 167–82.

Carbaugh, D. (2007b) "Quoting, "the Environment": Touchstones on Earth," *Environmental Communication: A Journal of Culture and Nature*, 1: 164–73.

Castro, J., Siebert, M., Hoffmann, J., Parks, M. M. and Thomas, M. (2015) "When I Say Nature, I Mean…" Performance presented at Bridging Divides: Spaces of Scholarship and Practice in Environmental Communication, June 11–14, Conference on Communication and Environment, Boulder.

Cox, R. (2007) "Nature's 'crisis disciplines': Does environmental communication have an ethical duty?" *Environmental Communication: A Journal of Culture and Nature*, 1(1): 5–20.

Elbow, P. (1989) "Toward a phenomenology of freewriting," *Journal of Basic Writing*, 8: 242–71.

Fassbinder, S. D., Nocella, A. J. and Kahn, R. eds. (2012) *Greening the academy: Ecopedagogy through the liberal arts*, Rotterdam, Sense Publishers.

Freire, P. (1970) *Pedagogy of the oppressed*, Continuum, New York.

Foucault, M. (1977) *Discipline and punish: The birth of the prison*, Pantheon Books, New York.

Greenwood, D. L. and Levin, M. (2005) "Reform of the social sciences and of universities through action research," in Denzin, N. and Lincoln, Y., eds., *SAGE handbook of qualitative research* (3rd edn.), Sage Publications, Los Angeles, pp. 43–64.

hooks, b. (1994) *Teaching to transgress: Education as the practice of freedom*, Routledge, New York.

Kahn, R. (2008) "From education for sustainable development to ecopedagogy: Sustaining capitalism or sustaining life?," *Green Theory & Praxis: The Journal of Ecopedagogy*, 4(1): 1–14.

Lasn, K. (2000) *Culture jam: How to reverse America's suicidal consumer binge—And why we must*, HarperCollins, New York.

Mignolo, W. D. (2007) "Delinking," *Cultural Studies*, 21: 449–514.

Milstein, T. (2011) "Nature identification: The power of pointing and naming," *Environmental Communication: A Journal of Culture and Nature*, 5(1): 3–24.

Milstein, T. (2012a) "Banging on the divide: Cultural reflection and refraction at the zoo," in Plec, E., ed., *Perspectives on human-animal interaction: Internatural communication*, Routledge, London, pp. 162–81.

Milstein, T. (2012b) "Greening communication," in Fassbinder, S., Nocella, A. J. and Kahn, R., eds., *Greening the academy: Ecopedagogy through the liberal Arts*, Rotterdam: Sense Publishers, pp. 161–74.

Milstein, T. (2012c) "Survive, critique, and create: Guiding radical pedagogy and critical public scholarship with the discursive guideposts of ecopedagogy," *Green Theory and Praxis Journal*, 6(1): 3–16.

Milstein, T. (2015) "The Inside-Out Classroom." Presidential Teaching Fellow Award acceptance speech presented at the Center for Teaching Excellence Awards Ceremony, University of New Mexico. Available at: http://publichealthnm.org/2015/S05/01/professor-calls-for-unm-to-divest-from-fossil-fuels/ (Speech recording) (accessed October 20, 2016).

Milstein, T. (2016) "The performer metaphor: "Mother nature never gives us the same show twice," *Environmental Communication*, 10(2): 227–48.

Milstein, T. and Pulos, A. (2015) "Culture jam pedagogy and practice: Relocating culture by staying on one's toes," *Communication, Culture & Critique*, 8(3): 395–413.

Moore, J. W. (2015) *Capitalism in the web of life: Ecology and the accumulation of capital*, Verso Books, New York.

O'Sullivan E. (2002) "The project and vision of transformative education: Integral transformative learning," in O'Sullivan, E., Morrell, A. and O'Connor, M. A., eds., *Expanding the boundaries of transformative learning: Essays on theory and praxis*, Palgrave, New York, pp. 1–12.

Plec, E., ed. (2012) *Perspectives on human–animal interaction: Internatural communication*, Routledge, London.

Rogers, R. A. (1998) "Overcoming the objectification of nature in constitutive theories: Toward a transhuman, materialist theory of communication," *Western Journal of Communication*, 62(3): 244–72.

Tucker, B. (2012) "The flipped classroom: Online instruction at home frees class time for learning," *Education Next*, 12(1): 82–3.

Part II

Diverse practices in teaching environmental communication

This part is broader in scope than Part I. Authors engage specific and diverse teaching approaches and emphases.

5 The role of social constructionism as a reflexive tool in environmental communication education

Lars Hallgren

Imagine the following critical situation arising in your first job. In one corner stands stakeholder A, who very strongly believes X about a resource R. In the other corner stands stakeholder B, who very strongly believes Y, which is the opposite of the belief X. The task is now to facilitate a constructive dialogue between A and B about R. Imagine another situation. Organization C wants the residents in village Z to commute to work by bus or bike instead of by car. You are employed by the organization to suggest measures to generate such a change.

Both cases describe typical work situations for graduate students in the Environmental Communication and Management (ECM) Master's program at the Swedish University of Agricultural Sciences (SLU). These situations are different and require various types of knowledge and skills from the professional actor. Many diverse theories, methods, and models could explain these situations. However, one core competence essential in both situations is the ability to reflect on the communication situation in which one participates. Planning ahead for communication situations and reflecting on and evaluating communication situations in retrospect are important. However, the ability to reflect on and respond to the dynamics of communication situations in which one participates *as they evolve* is equally important. This core competence includes being able to see and to understand how one influences and is influenced by the situation.

Environmental communication comprises all meaning constructing processes that emerge when humans interact with their surroundings and are attentive to the environment, potential resources, and issues of sustainability and unsustainability. In this chapter, as in the ECM education at SLU, the "environmental communicator" is a person who is professionally involved in these meaning constructing processes.

This chapter examines how social constructionism and symbolic interactionism contribute to the development of such reflective and meta-reflective competence in environmental communication education, why it is important, and how this perspective guides us when coordinating learning situations in the two-year ECM program on which I have been working since 2007.

Reflective competence in environmental communication situations

In our approach to education, the ability to reflect and meta-reflect on the communication processes that one is part of is an important competence for everyone working in environmental communication situations. This section discusses what is meant by reflection and meta-reflection, and why I consider it to be important for actors working in environmental communication situations. While I argue that reflection on communication is a core competence, I do not claim that it is the only important competence for environmental communicators.

Reflection occurs when actors involved in a sequence of interactions ask themselves questions about:

- how they understand the situation;
- how other actors understand the situation;
- how they and their actions contribute to others' understanding of the situation;
- how others' actions contribute to their understanding of the situation; and
- if there are alternative ways for them to interpret and to act in the situation.

Meta-reflection includes the reflection process itself in the reflection; that is, it involves the actors considering how their reflection and means for reflection are affecting the interaction between themselves and other actors. Communication refers to intersubjective meaning making, that is, social interaction where meaning is co-constructed. This view on social interaction is of relevance to the management of environmental and natural resources.

Social constructionism as a reflection tool

Social constructionism states that human beings' actions and their interpretation of their situation, including themselves, are dependent on other actors' actions and their interpretation of their situation. This dependency is mediated by symbols (such as words) and social objects, which are both a product as well as the context of social interaction (Mead 1934; Blumer 1969; Berger and Luckmann 1966; Garfinkel 1967; Linell 2009). Throughout this chapter, both social constructionism and symbolic interactionism provide the assumptive foundation to describe the meta-reflective pedagogical process used in this program.

Social constructionism explains the relationship between the individual and society and conceptualizes communication, intersubjective meaning making, as the core process in the constitution of society. However, in the present educational context, social constructionism and symbolic interactionism are not simply limited to a theory that students read, discuss, and apply when making an external analysis of society and communication; they also offer a framework for reflection and meta-reflection for those involved in environmental communication.

Social constructionism is relevant as a reflexive tool partly because it is explicit about reflection and gives self-reflection an explicit role in human action. Social constructionism, as outlined by Berger and Luckmann (1966), is closely aligned with the central tenets of symbolic interactionism as articulated by Mead (1934) and Blumer (1969). The principle idea in both approaches is that our social worlds are constructed through interaction. Both approaches are used throughout this chapter to articulate the educational process used in the ECM program. According to Mead (1934) and Blumer (1969), individuals become aware of themselves through social interaction with others: "the human being is an object to himself … this gives him the means to interact with himself … through further interaction with himself he may judge, analyze and evaluate the things he has designated to himself" (ibid., 62). Through observing others' reactions toward oneself, an individual constitutes his- or herself as an object for observation. This process takes place intersubjectively: "Interaction is … centrally a matter of negotiating identities and roles" (McCall 2003, 329). Through establishing oneself as an object, internal reflection, whereby individuals analyze themselves, becomes possible and provides the agent with agency (Blumer 1969, 63).

To avoid misinterpretation, an important distinction needs to be made at this point. As Berger and Luckmann (1966, 1–2) carefully point out in their discussion, social constructionism is not a contribution to a philosophical discussion on ontology; it is not about the character of reality but is a contribution to a discussion about action theory and epistemology. In the context of environmental communication research social constructionism would, for example, legitimize dismissing climate change as "only a social construction." However, how such a conclusion could be reached is unclear. The theory as it is applied here is purely epistemological and not ontological; it does not claim to explain the character of reality but can be used in an analysis of how human beings interpret different situations and act accordingly. The theory can contribute an understanding of what takes place in conversation between actors with diverging views on causes and consequences of climate change and how these different opinions have emerged. However, it does not make any statements about climate change or any other phenomenon beyond the intersubjective, language mediated, sense-making processes through which phenomena become part of society. For participants in a conversation on climate change, social constructionism theory can be a helpful reflection tool to understand how and why their statements may be misinterpreted by other actors and why they react in the way they do. This reflexive and meta-reflexive competence is a key factor for constructive dialogue in contested environmental issues.

As a tool for reflection, social constructionism directs attention toward the intersubjectivity that arises between actors. It recognizes that the meaning of an action, such as a gesture or a statement, is not entirely decided by the actor or by the recipient of the action but by the interface between the intention of the actor and the interpretation and response of the recipient. Observing the gesture of the other, or the consequences of that gesture, "affords me access to his subjectivity" (Berger and Luckmann 1966, 34).

Another important issue in social constructionism is recognizing that communication is a process of both integration and differentiation between actors. For communication to take place, commonality, joint symbols, and expectations are necessary as a minimum (Rommetviet 1974; McCall 2003; Linell 2009). At the same time, differences in perspectives are necessary; otherwise there is nothing to talk about. We communicate because we assume the other as "other" (thus different from us), and we communicate because we believe we have enough in common with the other to succeed in taking the perspective of the other. Similarly, the consequence of communication is commonality: shared meaning, concepts, and social objects. However, to identify the difference between self and other, we need to communicate.

A social constructivist lens on environmental communication situations

Environmental communication takes place in a large and varied number of social situations, situations that all differ in character. In these situations it is beneficial for the actors involved to reflect and meta-reflect. To demonstrate how social constructionism offers the environmental communicator a relevant language for reflection, I will return to the situations described in the beginning and briefly describe how they can be understood through a social constructivist language.

In the first situation the environmental communicator, let us name her Eustacia, is commissioned to facilitate dialogue between stakeholder Anna and stakeholder Bertrand, who both disagree about the conditions for a particular resource, a river, for example. A social constructivist language for reflection makes it possible for Eustacia to talk with herself about the situation and her own influence on the situation. For example, it is possible that Anna and Bertrand have different perspectives on the resource and are accordingly defining the situation and the resource in different ways. It might also be that Anna and Bertrand make sense of their disagreement in different ways and give different meaning to each other. Further, Anna and Bertrand have expectations toward the behavior of Eustacia that are different from Eustacia's own expectations toward her own behavior. When Eustacia is considering these differences in definition of the situation, she understands they will run a risk of misinterpreting each other's intentions. Being aware of the differences in perspectives and definitions of social objects, the self-reflecting Eustacia, through the language of social constructivism, will, in the dialogue, be able to explicate some of these differences in meta-communicative turns, and through that make it easier for Anna and Bertrand to investigate what they are disagreeing about.

It is quite obvious that reflective and meta-reflective competence is important for environmental communicators working in face-to-face interactions. However, reflective competence is also important for environmental communicators working with communication mediated by technology (communication strategies, media campaigns, advertisements). A strong reflective and meta-reflective competence might be the only reminder to the environmental communicator that he or she is involved in a process of co-constructing meaning, what Mead (1934, 253) called participation in the mind of the other, and not in a solipsistic, independent production and transmission of a message.

Meta-reflection inside and outside the ECM classroom

A fundamental component in developing constructionist reflexive competence is for students and lecturers to read and talk about social constructionist theory. However, that in itself will not lead to meta-reflexive competence. Take, for example, Mead's idea that self is not an intrapsychic structure, but a social process, and that self is constructed intersubjectively through social interaction. Being able to describe this abstract theory is wholly different from putting it to use in the midst of an environmental communication situation as one is reflecting on one's own and other's actions and the relationship between these actions as they are presently happening.

Students may have read and understood the important discussion by Ivana Markova (1990) about the triadic character of the communicative turn, but they will not automatically apply that understanding as a meta-reflective tool when making sense of a concrete environmental communication situation. Markova's theory becomes a tool for meta-reflection as part of the actors' repertoire as students acknowledge that the meaning of their own action in this local context also depends on the action they are responding to and the counteraction that in turn will follow. To develop this competence students need practice using social constructivist language for talking through everyday situations.

When lecturing about specific environmental communicative phenomena such as democracy, deliberation, disagreement, conflict, power relations, self-processes and the concept of symbolic interactionism, we consistently apply these kinds of phenomena as they appear in the classroom and make sense of them through social constructionism terminology. The aim is to demonstrate that the somewhat abstract terminology of the theory is applicable for reflecting on concrete everyday situations.

One guiding principle when designing learning situations is that learning about meta-reflective practice demands different levels of distance and involvement, respectively, to such situations. It is sometimes easier to apply theoretical tools when analyzing situations where the analyst is not involved, while increased involvement often results in more internal analyses on one hand and less critical analyses on the other; the strength of the theory is exchanged with commonsense explanations. However, external analysis of a situation refers to situations where the analyst is not involved and practicing such an analysis does not necessarily result in a meta-reflective competence, which is our aim. I believe the solution is to alternate between analysis of different levels of distance/involvement and between situations of different levels of importance for the analyst/student.

I present here a number of approaches and exercises we have devised within the ECM program to develop meta-reflexive competence based in social constructionism:

- *"What is the color of snow?"*
  This exercise illustrates the constructing connection between language, perspective, perception, and thinking. The exercise takes less than five minutes to complete and fits well into more theoretical lectures about fundamental social constructionism, the perspectivity of reality perceptions, and the role of language and communication in social theory. The lecturer shows a picture of a snowy landscape and repeatedly asks the class, "What is the color of snow?" The question is repeated six to eight times, until all students answer "white" in chorus. Without warning the question will be changed to "What do cows drink?" Normally at least two-thirds of the group answer, in chorus "Milk!" and then "Oh, no, water!" The lecturer asks the students what they think happened, and then explains this through social constructionism: the meaning we make by using symbols is confirmed and reconstructed each time we successfully use these symbols to describe our reality by talking about it. In this case there has occurred a confusion of symbols because of how they connect to each other in different contexts. We experience with the help of language and this language biases our perception. There is no other way for humans to perceive the world than through organizing experience with the help of socially agreed categories that are constructed and reconstructed through their use in social interaction.

- *Interpreting pictures, reflecting and meta-reflecting on interpretation.*
  The aim of this exercise is to understand how meaning and sense-making function as social processes through experiencing, observing, and reflecting on the process of interpretation; that is, how humans give meaning to objects and situations.
  In small groups students discuss what they see in a photo provided by the lecturer and prepare a rich and detailed story about it. In the discussion the lecturer acknowledges how language, social norms, and conceptual models have influenced the narrative made up by the students. A typical dialog between students and lecturer could look like:

STUDENTS: In this picture we see workers in a factory and the factory is polluting the air.

LECTURER: Why do you think they are workers in a factory, and why do you think the factory is polluting the air?

STUDENTS: Because they wear safety helmets and there is dark smoke coming out of the chimney of the factory.

LECTURER: So you came to this conclusion because you are comparing what you see in the picture with a reference object of which you have experience and which is represented by language; the words "safety helmets" and even "dark smoke" make it possible to connect this picture with a meaning that you have experienced in another social situation.

Note that the picture carries meaning as does the socially agreed language we use for talking about the picture. This exercise also provides an opportunity to reflect about how a critical, theory driven awareness affects the process of sense making.

- *Students draw pictures*.
  The aim of this exercise is to acknowledge the importance of language in the experience of social objects and to show how language perpetuates ideas about nature. It also highlights the process of action, the iteration between dialog with self and action. The students make a drawing of their own hand and discuss afterwards how they did it. The lecturer supplements the students' descriptions by using the language of symbolic interactionism and social constructionism:

  a We plan our action: look at the motive, plan to draw a line on the paper;
  b we act, manipulate, and intervene with our context: draw a line;
  c we observe the consequences of our action, considering the meaning of our action and the need for altering action.

  The interaction with a paper by using a pencil is to some extent comparable to the interaction with others through the help of symbols. The vital difference is that, as opposed to the visible traces left by a pencil, an oral interaction does not leave any concrete traces of action except as an invisible intersubjective context for the proceeding action. This is the same sequence of action we can experience when we are involved in a conversation or a conflict about the environment. Namely, actors act toward each other to recognize, make sense of, and reflect on the construction of meaning that the response of the other constitutes; they alter their actions accordingly. Further, when we, who are not trained artists, draw, for example, a hand, we tend to draw in accordance with language-based categories: we draw a generalized thumb, generalized palm, and so on, and these generalized social objects are not necessarily similar to the shape of the hand we have in front of us when drawing. The language we are using for naming a hand is directing our attention. Language steers attention toward social objects as already being predefined in language. This is, of course, also an important aspect of environmental management.

- *Analysis of dialogue transcripts*.
  In this exercise we provide a video recording of an environment-related public meeting and a transcript of the conversation. Students are asked to gather in groups of three to analyze how utterances relate to one another and to notice meta-communicative turns. The lecturer also presents a pre-prepared analysis based on social constructive concepts and theory, and lecturer and students discuss and compare their analysis. The focus is on how actions are both dependent on the local communication context and

generate communicative context for future action. If students suggest an analysis based in personality theory, or if they focus on whether one argument is truer than the other, the lecturer suggests returning to interactionist theory. The lecturer redirects student focus on understanding how the argumentation is dependent on the dynamics of the interaction and generating context for future action as the important question. This is an example of an exercise where the analysis is performed on a situation that is external and distant to the students. To connect to their experience, students are also asked to imagine how they would react if they were to switch places with the participants in the environmental communication situation.

- *Group relations conference in micro format.*
  This exercise is inspired by the group relations conference of the Tavistock tradition (Miller 2004; Long 2004). However, instead of the 3–14 days of conference with reflection on group dynamics used in the Tavistock tradition, our exercise is limited to 45 minutes and, rather than a psychoanalytical focus, we seek to make sense of the experience with the help of social constructionism. Before the students arrive in the lecture theatre, brief written instructions are placed on all seats in the room. The instructions state that the lecture starts at 9:15 a.m. (which is in accordance with the timetable) and that the time between 9:15 a.m. and 10:00 a.m. is dedicated to investigating the social dynamic of the social system in the room and how it emerges, using the means and methods all students find useful for the task, without transgressing Swedish laws. Furthermore, it is suggested that attention of the investigation be directed to boundaries, mandates, and expectations that are constructed in this social system and its subsystem, and that participants take notes during the exercise. At 9:15 a.m. the lecturer starts making notes and observing what is happening but takes no other initiative to point out that the exercise has started. It usually takes some time before students realize that the lecture has started and that there will be no additional instructions. Once this happens, it often becomes quiet for some 5 to 15 minutes while the students repeatedly read through the instructions. After some time, some participants start talking in pairs, initially whispering. Occasionally, students try to take initiatives to organize joint action, create bigger groups, write things on the whiteboard, etc. The instructions for the lecturer require him or her to participate and support the investigation of the dynamic of the social system while it is emerging. However, the lecturer ignores all initiatives directed towards him or her, such as students asking questions relating to other issues, as for instance, "When is the deadline for the exam?" or "What should we do?" Thus, the lecturer should exercise no authority beyond the written instructions. This is a very confusing and sometimes tense exercise, stressful for both students and lecturer, but it makes it possible to observe and reflect on power relations, social norms, group formation, how mandates are related to expectations and norms, and how these phenomena are constructed and

maintained through social interaction. When the 45 minutes have passed, the lecturer signals very explicitly and warmly that the exercise is over and that he or she will again take on the authoritative role of being the responsive lecturer. The students are now asked to describe their experiences of the exercise, what norms they perceived, and how these norms were created and maintained by the participants. Often the talk is directed to the expectations participants have about what will happen in social situations, such as a lecture; how these expectations guide our actions and, when they are not met, generate insecurity. Furthermore, in the discussion we recognize that the same social dynamic is playing a role when people come together to coordinate actions on the environment.

- *Environmental and natural resource management conflict project work.*
  This is an exercise in which students work in project teams for three and a half weeks, one week of which is fieldwork. The project aims are twofold:
  - Students plan, conduct, and present an investigation of a local environmental or natural resource management conflict, and determine how the actors involved in the conflict perceive the situation.
  - They also observe and reflect on how they interact within and outside the project team when planning and conducting fieldwork and analyzing their material.

  How do they make decisions? How do they coordinate action? How do they handle misinterpretations? The project must be presented in two separate reports, one describing the investigated ENRM conflict situation, and one describing the group process of the project team. The teaching team offers supervision in both tasks. In the investigation task, the project teams ask for advice about design of interview situations and questions. The supervisors try to avoid direct instructions. Instead they act as facilitators, helping the project team to imagine how their intervention will affect the interview participants and to consider to what extent it is in line with what they want to do. Occasionally, project teams ask supervisors for consultations when they have disagreements, separations, subgroupings, or conflicts. The supervisors then use different methods for facilitating reflexive conversations between project team members, so they understand the process they are part of and how the different team members have interpreted the situation. In all consultations, the supervisors include a meta-reflection, where supervisors and project team members reflect on how they think they are affected by the current consultation situation.

Reflection on methods for reflection

Above I have described six exercises that we use or have used in the ECM program and that we believe contribute to developing competence in using social constructionism for reflection on environmental communication situations. These exercises are continuously reshaped, reviewed, and replaced. As teachers we also reflect and meta-reflect on what social interaction we are involved in, how we affect and are affected by other participants in these processes, and how our interventions affect other participants' ability to learn and reflect. Moreover, the theories we have proposed are in themselves social objects, directing the attention of the observer, making aspects of reality visible or hidden. However, these exercises illustrate what we are working on: generating situations to reflect on, suggesting a theoretical language for reflection, making reflection through dialog, creating space for meta-reflection, and connecting experience of reflection in the classroom to reflection in environmental communication action situations. One of the students who graduated from the ECM program in 2015, wrote the following:

> I explored Symbolic Interactionism with its different levels and developments and tried to completely understand it. My acquired knowledge of the theory helps me now to apply it in my everyday life. I can rethink social processes and explain them to myself. I can understand relationships better, as well as the communication between individuals. And, most of all, I can reflect on a deeper level on my own behavior. My whole individual process with Symbolic Interactionism showed me for the first time the applicability of theories in the "real world" and their usefulness not just necessarily on a big scale such as a research project but also for an individual in personal life.

What she describes is one of the competencies I believe to be essential for anyone working in any form of environmental communication situations. In their group process reflection report, one project team from 2015 wrote:

> In conclusion, we think this assignment was a great opportunity for us to learn and reflect over how we as individuals and as a group function. This knowledge and experience will help us further on in both our professional and social interactions.

Another project team from 2015 noted:

> Actually, we feel like we learned more from reflecting on our group process than from the case study, therefore it has been very useful to put theory into practice. The whole process has also proven the importance of effective communication, trust building, and time management.

Students graduating from the ECM program have become employed as practitioners responsible for the coordination of environmental management in public and private organizations in countries such as Sweden, Oman, Ghana, Cameroon, and New Zealand. Some of them work designing and conducting information campaigns. Some of them work with conflict management and facilitating dialog about contested environmental issues. The two situations described at the beginning of this chapter are typical in their professional practice. When standing there in the middle of the stream of action, they will remember "What does the cow drink?" and know that language influences attention and that certain aspects of social objects will be acknowledged by the language used, while other aspects are hidden. They will recall the group relational conference and know that socially constructed expectations and roles are important for the internal and external dynamic of an organization working with environmental change. They will remember drawing their hand and think of the iteration between action and reflection and its importance for understanding how actors involved in conflict are interpreting the action of their adversaries and how they adapt their own means for influencing the process to their interpretation of others' actions. Moreover, social constructionism offers a language enabling them to talk about their experience of interacting with others in the process of co-constructing meaning. The exercises I have described here have, in combination with reading classic social constructionism literature, facilitated the development for a fluent use of that language.

References

Berger, P. L. and Luckmann, T. (1966) *The social construction of reality: A treatise in the sociology of knowledge*, Penguin, Harmondsworth.

Blumer, H. (1969) *Symbolic interactionism: Perspective and method*, University of California Press, Oakland.

Garfinkel, H. (1967) *Studies in ethnomethodology*, Blackwell, Oxford.

Linell, P. (2009) *Rethinking language, mind, and world dialogically: Interactional and contextual theories of human sense-making*, Information Age Publishing, Charlotte.

Long, S. (2004) "Building an institution for experiential learning," in Gould, L. J., Stapely, L. F. and Stein, M., eds., *Experiential learning in organizations: Applications of the Tavistock group relations approach*, Karnac, London, pp. 101–36.

McCall, G. J. (2003) "Interaction," in Reynolds, L. T. and Herman, N. J., eds., *Handbook of symbolic interactionism*, Alta Mira Press, Walnut Creek, pp. 327–48.

Markova, I. (1990) "A three-step process as a unit of analysis in dialogue," in Markova, I. and Foppa, K., eds., *The dynamics of dialogue*, Harvester Wheatsheaf, Herefordshire, pp. 129–46.

Mead, G. H. (1934) *Mind, self, society: From the standpoint of a social behaviorist*, University of Chicago Press, Chicago.

Miller, E. J. (2004) "The 'Leister' model revisited," in Gould, L. J., Stapely, L. F. and Stein, M., eds., *Experiential learning in organizations: Applications of the Tavistock group relations approach*, Karnac, London, pp. 11–18.

Rommetviet, R., ed. (1974) *On message structure*, Wiley, London.

6 "Deep impressions"

The promise and possibilities of ecocultural experiential learning

Aaron T. Phillips

The title of this essay comes from a piece of reflective writing by a Chinese student in her first academic year at the University of Utah in which she considers her experience in a course titled "Tourism in the National Parks: Yellowstone." The student was reflecting on a four-day tour through this iconic landscape, an experience that left a "deep impression" (student response). Experiences such as the one this student had in Yellowstone National Park offer an immersive element that mobilizes meaningful engagement with a learning cohort (inclusive of instructors and students), with landscape, and with language. Taking place in an iconic landscape that students had not experienced before, this trip provided a unique opportunity for studying interactions with landscape through language. Language has a fundamental "participatory proclivity" (Abram 1996, 138) with landscape, even if at times it may be a tool of abstraction that can dilute experience (Grigg 1994; Halliday 2001; Stibbe 2010). Though this abstraction from immediate experience may be a feature of language, this study nevertheless uses language—and writing in particular—as a means of investigating how connections with place, with nature and with (eco) culture unfold. Overall, the interactions during the course seemed to catalyze cultural and linguistic exchange while providing opportunities for engagement with place and with landscape.

This chapter uses reflective writing samples and participant observation techniques to analyze intercultural communication through experiential learning in a diverse and largely L2 (multilingual students for whom English is not a first language) group of thirty students (77 percent Chinese, 13 percent U.S., 7 percent Japanese, 3 percent Congolese) in a Parks, Recreation and Tourism course. The course attracted a large number of L2 students due to a partnership between the University's Department of Parks, Recreation and Tourism and its English Language Institute, which offers an intensive preparatory English language program for international students. In this chapter, I present and analyze students' experiences of this field session, as recorded in their writing and other interactions. I suggest that a field-based method of engaging language and landscape might foster humanature connections, facilitate language learning, and catalyze cultural exchange. Further, the results of this study suggest that a course need not be explicitly focused on language acquisition/development or on environmental communication to exert powerful effects on students' environmental and linguistic understanding. Instead, connections with landscape, culture, and language seem to seed spontaneously in the rich interactive zone of the experiential learning context.

This chapter is conceptually framed by recent scholarly investigations of the many challenges and opportunities of ecocultural communication, particularly in pedagogical settings (Dickinson 2011, 2013; Milstein 2012; Milstein and Pulos 2015), and by the prominent role place plays in all communicative endeavors (Carbaugh 1996; Carbaugh and Cerulli 2012). In considering this course as a case study in emplaced ecocultural communication, I unite disparate threads of academic inquiry in the service of investigating the research question, "How does an experiential learning course in Yellowstone National Park affect L2 students' attitudes toward communication/writing in English?" Ultimately, I seek to put into conversation valuable strands of thinking and practice in ecocultural communication, experiential learning, and theories of L2 communication and writing.

Communication as an ecocultural phenomenon

Given this chapter's foregrounding of experience in an iconic landscape often associated with experiences of "nature," I place the chapter in conversation with recent advances in theorizing ecocultural communication that bring communication out of the strictly human framework often implied by intercultural communication scholarship. An ecocultural worldview has been advanced as a means of interrogating the persistent cleft between that which is "nature" and that which is "culture" (Dickinson 2011), particularly as it pertains to scholarly inquiries of culture and communication (Milstein and Pulos 2015). The compound terms ecoculture and ecocultural represent critical reappraisal of the Western mind/body split and its concomitant splitting of humans (in a Cartesian "cogito" formulation, the sole thinking and be-ing thing in the lifeworld) and nonhumans (ibid.). I apply the notions of ecoculture and ecocultural interaction in the context of this study, as they pertain directly to the intercultural exchange inherent in the trip and its function as an opportunity for course participants (students and teachers alike) to engage interactively with nature in ways that may move past human-nature dualism. I consider "nature" an experiential field both inclusive of and beyond humans (Phillips 2014), a notion advanced with linguistic certainty by the compound term "humanature" (Milstein 2011; Milstein and Kroløkke 2012).

Emplaced experiential education

Communication is thoroughly emplaced (Carbaugh 1996), which is to say it is seated in identity, affect, and culture (Carbaugh and Cerulli 2012). Just as it is emplaced and cultural, communication is inflected by experience (the immediate, felt experiences of the senses and the experiences of perception, cognition, affect and interaction) and connected to identity. This chapter allows an interface with identity, difference, and with cultural ways of knowing/being through the practice of experiential learning. I apply to this chapter Chickering's (1976, 63) definition of experiential learning as that which "occurs when changes in judgments, feelings, knowledge or skills result for a particular person from living through an event."

I use the term experiential learning as distinct from experiential education, emphasizing the living, learning subject over the subject being taught. Such an approach is in keeping with the primary philosophical underpinnings of experiential education as expressed by influential educational theorists as Dewey (1916) and Hahn, the founder of Outward Bound, a pioneering presence in experiential education (James 1995). For Dewey, an educational program centered on subjective experience foregrounds the learner's reflections on experience as a powerful learning tool; for Hahn, experiential learning must be centered in experience and rooted in socio-cultural processes (Smith and Knapp 2009). Underlying experiential learning, of course, is the felt experience of the phenomenal world. As Abram (1996) has shown, linguistic experience and perception are thoroughly entwined. I proceed, then, from the notion that culture, place, felt experience, cognition, affect, identity, and language combine in our experience of the lifeworld.

Interfacing with language and writing theory

Canagarajah (2002, 1) has argued that the work of language and culture learning is best accomplished in "a rich context comprising diverse social institutions and experiential domains." The study situation here is an attempt to embody such a domain for the purposes of inquiring how such an experience affects students' sense of language. Since a portion of the chapter material is formed of students' written responses to their experience, the chapter interfaces with the field of writing studies in which theorists have long posited ecological models of literacy and composition (Cooper 1986; Dobrin and Weisser 2002; Weisser and Dobrin 2001). Cooper's (1986) theory of the ecology of writing extended discursive practices beyond human language, as had earlier studies in "ecolinguistics" (e.g., Sarles 1977). These formulations argue—and I agree—that human linguistic practices are fundamentally coextensive with the worldwide web of the "ecologies in which we swim" (Condit 2006, 370).

Methods

I used student writing samples and my experience as a participant observer in the course to analyze students' experiences. During the field session of the course, I took field notes and collected three writing samples over a four-day period on a field trip to Yellowstone National Park on October 8–11, 2012. The field trip was part of a non-graded, credit/no-credit, two semester-hour course in the Department of Parks, Recreation and Tourism at the University of Utah entitled "Tourism in the National Parks." I participated in the field trip as an instructor and communication/writing specialist. Based at a hotel outside the park, we ventured into the park each day to take part in hikes to various park attractions, listen to interpretive talks by rangers, and visit park facilities such as museums and visitor centers.

Student reflective writing

Students in the course engaged in a pre-trip writing exercise in which they identified their expectations of the trip; a reflective essay after the second day of the trip about what experience from that day stood out most in their memory and why; and a post-trip writing assignment. The post-trip writing asked students which moments in the trip as a whole they found most memorable and why, along with whether they thought the trip helped them develop communication skills. This question about communication was purposefully open-ended. That is, it did not isolate a particular aspect of communication in an effort to see what patterns might spontaneously emerge in students' appraisal of their communication skills in connection with the course. Combining my experience of the course with analysis of students' written responses to the course, several compelling patterns regarding connections to landscape and language emerged. For the purposes of this analysis, these patterns are threefold:

- The course opened up communicative channels between human and nature.
- The course seemed to facilitate language learning.
- The course seemed to catalyze cultural exchange.

Importantly, these patterns are prominent upon analysis even though the course was designed as a tour of a National Park to learn about land management strategies, not necessarily to learn communicative practices. This suggests that something about the experiential format of the course may be a primary factor in building connections between landscape and language.

Building experiential humanature connections

From my perspective as an observer, a general mood of curiosity and openness characterized almost all individual members of the class and of the class as an aggregate. During the pre-trip meeting, I noticed a palpable air of excitement in the classroom, made evident by a flurry of questions on what kind of weather conditions to expect, what kind of wildlife students might encounter, and what equipment to bring on the trip to ensure a comfortable experience. This anticipatory group energy, fueled by excitement about the unknowns ahead, seemed to manifest not as a negative, fear-based group atmosphere, but rather as a positive orientation toward the exploratory work of travel and learning. During the trip, curiosity naturally waxed and waned, but piqued around certain activities, most notably visiting thermal features in the park, the risk/excitement/reward associated with the activity of hiking, and experiencing a quiet moment of technology-free reflection near Trout Lake in the eastern portion of the park. Here are some notable excerpts quoted from the students' mid-trip essays that speak to the attitude of curiosity and excitement that characterized the collective experience:

CHINESE STUDENT: I have been told that in the red and yellow areas [of thermal features], there are a lot of bacterium. I am really surprised that bacterium are colorful and can survive in such high temperatures.

CHINESE STUDENT: When we were separated into many groups, we enjoyed a really excited hiking activity. Because today was my first time to pass through the very cliffy road since I was 12 years old, it was really adventurous.

CONGOLESE STUDENT (L1 FRENCH): The most memorable part of the trip was the 3 min. exercise of silence to listen to nature. Why?

- Very relaxing
- Helped me focus
- Helped clear my mind and listen to the sound of nature
- Learned for the first time how the wind blows.

These three responses are connected by a shared attitude of openness to the "new experiential domains" Canagarajah (2006) advocates as ways to involve the learner in an experiential enmeshment with language. The Congolese student spoke with the class after the evening's reflective writing, saying the silent moment (an admittedly manufactured silence) allowed her to hear, for the first time in her life, the sound of wind blowing through stands of trees. This new experience was mediated by language—the student reflected on the experience orally and in writing—but was also phenomenologically vivid and sensory. Without an attitude of curiosity about the sound of the wind, "extremophile" bacteria that inhabit a hot spring, or what lies beyond the next hill, these students would not have learned and changed as they adapted to new ways of being in the world phenomenologically, linguistically and, ultimately, culturally. For these students, the events of the trip mobilized curiosity and engagement with the experiential domain in new ways. From my perspective, these connections seemed to be palpable and genuine for students. Students' experience of enmeshment in the world around them may have fostered the kind of humanature connection that may flatten easy distinction between the experiencer and experienced.

Perhaps no experience calls into question our separation from the world in which we dwell quite so much as weather, particularly wind. We breathe in the air brought by the wind and exhale it back into the swirling currents around us as we become wind and wind becomes us. Students in the course reflected on this kind of connection. For example, the Congolese student expressed wonder at hearing the sound of the wind moving through the trees and feeling the wind blow. The student's reflection seemed to connect her to wind and language alike; that is, the novel experience of listening to wind for the first time and the experience of reflecting on that experience through language tied experience—perceptual, emplaced, affective—to the representation of that experience in a paradoxically direct way. Far from being a removed record of experience, writing about this situation seemed to cement the experiential–linguistic connection for this student. Feeling and hearing wind or other natural phenomena transports the experiencer to the roots of linguistic experience. Indeed, philosophers have long located the origins of language in the experience of the lifeword: Vico (1961 [1725]), for example, suggested language arose spontaneously out of expressions of wonder at or fear of natural phenomena. This student is writing in her second language to express an experience she had for the first time. Mapping a positive, wonder-filled and novel experience in one's second language may catalyze a positive orientation toward the new language though which one describes that novel experience. Since the use of writing in educational contexts is usually connected to testing in one form or another, this use of writing to recount a sensorially rich and unstructured activity using a mode of writing unattached to grading or testing is likely, I think, to foster a productive orientation toward the use of English and connect the use of English to elemental experience of the phenomenal world.

The student whose reflective writing discussed the hike engaged a central component of experiential learning: risk. Conscious risk-taking catalyzes the transformative element of experiential learning (Darling-Hammond 1998; Edge 2002). Risk can take many forms and does not of course always involve the palpable sensation of danger the student recounts. Yet the experience of risk followed by a subsequent negotiation through risk and into safety is illustrative of transformative, experiential, and emplaced learning. Fundamentally, this student learned that her limits may be more expansive than she may have thought when it comes to a fear of heights; this learning process was catalyzed by an experience of risk. Like the experience of listening to the wind, and perhaps even more so, this experience featured a visceral element that went to the core of felt experience. Recounting that experience through reflection in a second language may create a new affective orientation toward a second language by bringing it closer to the felt experience of risk and the overcoming of fear.

Emplaced experience and language

When the Union Pacific Railroad began offering regular service to Yellowstone National Park in 1925, its promotional materials gushed, "what distinguishes Yellowstone is the existing evidence of those original mysterious and magnificent processes of the Park's creation. The visitor walks through the looking glass into a wonderland where incredible things happen" (UPS 1925, 5). While these words marketing the Yellowstone experience to tourists nearly a century ago seem quaint now and clearly indicate the way the park has historically been and continues to be "sold" as a consumable commodity, the theme of wonder that characterizes the railroad's promotional material was echoed many times in student responses to the field trip. Students captured their wonder by reflecting on their experience in English. As I analyzed these responses from the mid-trip reflective writing, students' insights spoke to deep connection with place:

> CHINESE STUDENT: Geysers and hot springs are unforgettable to me. They are always steaming. Those areas are like a paradise in ancient Chinese legend. But sometimes they are just like battlefields full of smoke and gunpowder. And they always make boiling voice and they have different colors and shapes.

In this response, the student makes a specific connection to home culture, to language and to place. These connections are productive ecocultural fusions. The use of the term "voice" when discussing the geysers is interesting here, as is the paradise/battlefield binary the student presents. The phrase "boiling voice" is exceedingly rare (its only prominent appearance is in a Thomas Hardy poem), but for this student the phrase renders the language of the landscape. This uncommon collocation is thus perfectly suited to the particularity of the student's experience in that place, at that time, and through the hybrid language of home and afar. The reference to legend knits the student's experience of this place to that of home culture.

The next two snippets speak to the sublime experience, a human response to the lifeworld that, while not entirely unproblematic, may have potential to bridge the chasm between human and "beyond human" (Abram 1996):

> CHINESE STUDENT: Standing beside the steep and high cliff, I became forget all the unhappiness of school or even I have forgotten I am a human being, as I was totally intoxicated by the nature. The sound of water dropping from the high level; the wind blowing on my face; the sunshine surrounds me. All of these feelings are out of my surprise.
>
> CHINESE STUDENT: The scenery there was really beautiful, not only has the exquisite quality beauty, and has mountain and canyon's beautiful about the sturdiness, let a person enjoy them. Let the person has a kind of impulse living here.

These responses carry the tone of a sublime response to nature, a deep, emotional connection to the experience of self in place. The sublime as a Western cultural construction is perhaps most closely associated with the philosophy of Edmund Burke (1909). Christine Oravec (1982, 219) describes the emotional process of the sublime response; it is remarkably similar to the second student's response above:

> First, the individual is exposed to the sublime object—usually an outstanding natural object—which produces sensations of overwhelming magnitude and quality. The individual attempts to comprehend, intellectually and

> emotionally, the sublimity of the natural object, but failure to do so leads to the second stage of the response. Here a negative state is produced. Whether a result of fear, alienation, awe, or loss of identity, the sense of self is diminished and the individual feels a separation from, or lack of control of, the natural environment. Finally, the individual moves past the feeling of inferiority to a sensation of exaltation, or at least wonder, at the relative grandeur of the object compared to the self.

In some American cultural discourses, the sublime has been used rhetorically in "artistic and touristic" (DeLuka and Demo 2000, 249) discourses to build social capital: experiencing moments of the sublime through touring in natural areas or viewing art representing those areas can be seen as building cultural capital. Yet for the student who reflects on the experience of being "intoxicated" by nature, the reflection is not a simple adoption of Western cultural values; instead, it is an ecocultural synthesis of felt experience and knowledge of culture. Knowledge of culture, in this case, is an ecocultural fusion of both the Chinese esthetic construction of nature, which is often figured as "affinitive to" the sublime (Wong 1986, 125) and the Western notion of the sublime, which itself transacts on a global level with Chinese culture and vice-versa. These experiences of wonder and delight speak to an ecocultural connection between the experiencer and the experienced—a humanature fusion—rather than functioning solely as individual esthetic experiences. Adding to rather than merely recasting the experience itself, the reflective writing seems to fortify ecocultural connections to place and with the social scene of the unbounded experiential classroom. This immersion in felt experience, coupled with written reflection on that experience, binds the learner to place and forms bridges from home culture to other cultures, from person to place, and from home language to a new language.

Experiential interchange of culture and language

Just as the course offered opportunities for ecocultural experiences of the world beyond (and encompassing) humans, it also provided a fertile cross-cultural environment for interaction around language difference. As I have suggested, the experience of language is always rooted in the place in which it unfolds, and the linguistic interchanges the course fostered were complementary of the ecocultural fusion of the experiencer and that which was experienced. Interactions with language and difference, then, were an extension or outgrowth of the class's interface with nature. Each of the 30 students in the course brought their own culture, language and worldview to the class situation. The openness with which many students approached the experiential learning of the trip enabled a substantial interchange of knowledge across cultures and languages. This interchange often took a dynamic, exchange-based form involving the sharing of linguistic and cultural information. This exchange of experience and knowledge arose spontaneously among student groups. On the second day of the trip during the bus ride into the park, a particularly outgoing Chinese student declared aloud his goal of learning to count to ten in as many languages as possible. An Alaskan-born American student taught the student to count to ten in Spanish; the French speaker from Congo taught the student in French. This form of language interchange continued throughout the trip. Below are some student reflections on that interchange, excerpted from the post-trip writing:

CHINESE STUDENT: I met a lot of foreigner friends during the trip. We shared the feelings, experiences, and cultures together. On the other hand, our foreigner friends can also learn Chinese from us.

This student's reflection on shared experience takes us to the root of experiential learning and ecocultural exchange. On the course, there was no division between experiencing and sharing: the class experienced the park as a collective, and their experience of it was immediately yoked to the interchange of culture and language. In this excerpt, the student references language learning immediately after referencing "feelings, experiences, and cultures." The boundarylessness of this linkage between place, language and the interface of different cultures seemed to foster the creation of meaning, as the next excerpt implies:

CHINESE STUDENT: Listening to the instructor, and making friends with friends nationwide are both meaning for me. First of all, it is a good opportunity for me to practice English by communicating with others. What is more, learning other countries' languages is also interesting; in this way, we have come to know a lot of languages by using English, such as Spanish, Japanese, and also it is interesting for me to know others' culture, such as food, schools, hobbies."

Here, the student notes the practical efficacy of using English as a *lingua franca* to communicate with other students and with teachers. Food, a key marker of culture, also animates this student's response. On the trip, food was a constant source of discussion, a point of exchange where the differences between various cuisines and the similarity of basic tastes were brought up in unison. Food emerged as a key index of culture, place, and difference on this course, but it also brought the participants together in discussion. The presence of a bison burger on the menu at a West Yellowstone restaurant, for example, elicited discussion about native fauna, providing further connection between place, language, and experience.

Further responses about the course's effect on communication and culture showed a consistent trend toward place-based language and cultural exchange:

AMERICAN STUDENT: The time with the international students allowed me to work on my communication skills with Chinese students. I enjoyed helping the Chinese students understand general lingo words that we use in the United States. Another experience in working on communication skills with a Chinese student [name] was a good time. It was a good time because he wanted to learn other languages' basic words and how to count to ten. I thought it was amazing that a foreign student was so indulged into learning more about other languages than English.

JAPANESE STUDENT: I like to talk to people who have grown up in different background because they often told me what I have never expected.

CHINESE STUDENT: [Name] is a really thoughtful person. We had chatting with our major and after chatting I felt like I had a clear thought about my major. I was not so clear before. But during our conversation, I felt I need to study more hard. Anyway, it was a great experience talking with native speaker about my major.

CHINESE STUDENT (WITH BACCALAUREATE DEGREE IN ENGLISH FROM CHINA): I greatly appreciate all the communications with all of our team members, not only so many native speakers, but also students from Japan and Congo. I talked assorted topics with different instructors and students. I talked with Aaron about some novels and movies and other interesting story about the trip. I talked with [name] about philosophy, politics and other huge stuff. I talked with [names] about our biology, anthropology and exercise major a lot!! I felt huge improvements for my English expression and got more confidence and faiths!!

These responses reflect students' experience with the rich ecocultural interchange the experiential learning environment provided. The bus was a rich zone of interaction; activities such as hiking engaged the imagination, added an element of risk and provided further richness for ecocultural contact. The encompassing nature of the course—students ate together, traveled together, hiked together and roomed in pairs—provided a much higher-context experience than a temporally fractured sequence of one-hour course meetings characteristic of most university courses. Because of this, and perhaps also because of the non-graded nature of the course, students openly engaged one another on subjects that interested them. Many students seemed particularly interested in cultural and linguistic differences. Even though the course was not about learning other languages or cultures, the class group spontaneously came together around differences—and similarities—in language and culture. The literal translation of numbers from one language to another seemed to foster deeper new identifications with the interwoven ecologies of people, place, language, and landscape.

Conclusion

This productive and potentially transformative field session suggests student attitudes toward nature, culture, and communication are likely shaped in positive ways by experiential learning in a context like Yellowstone National Park. My experience of the field session and subsequent analysis of it both indicate the trip fostered humanature connections, facilitated language learning, and catalyzed cultural exchange. Ultimately, this experience seemed to bind all of us as learners to place and to one another in powerful ways. To explore these intriguing connections further, future studies might work more longitudinally, following up with students who have engaged in an experiential learning environment and considering how language acquisition and the development of ecocultural connections to place interact. Such research may seek to delve into further questions regarding the influence on second language acquisition experiential learning might yield. The terrain is certainly fertile.

References

Abram, D. (1996) *The spell of the sensuous: perception and language in a more-than human world*, Random House, New York.

Burke, E. (1909) *On the sublime and beautiful and reflections on the revolution in France*, P. F. Collister & Sons, New York.

Canagarajah, A. S. (2002) *Critical academic writing and multilingual students*, University of Michigan Press, Ann Arbor.

Canagarajah, A. S. (2006) "The place of world Englishes in composition: pluralization continued," *College Composition and Communication*, 57(4): 586–619.

Carbaugh, D. (1996) "Naturalizing communication and culture," in Cantrill, J. and Oravec, C., eds., *The symbolic earth: Discourse and our creation of the environment*, University Press of Kentucky, Lexington, pp. 38–57.

Carbaugh, D. and Cerulli, T. (2012) "Cultural discourses of dwelling: Investigating environmental communication as a place-based practice," *Environmental Communication*, 7(1) 4–23.

Chickering, A. (1976) "Developmental change as a major outcome," in Keeton, M., ed., *Experiential learning*, Jossey Bass, San Francisco, pp. 62–107.

Condit, C. (2006) "Contemporary rhetorical criticism: Diverse bodies learning new languages," *Rhetoric Review*, 25(4): 368–72.

Cooper, M. (1986) "The ecology of writing," *College English*, 48: 364–75.

Darling-Hammond, L. (1996) "The quiet revolution: rethinking teacher development," *Educational Leadership*, 53(6): 4–10.

DeLuca, K. and Demo, A. (2000) "Imaging nature: Watkins, Yosemite, and the birth of environmentalism," *Critical Studies In Media Communication*, 17(3): 241.

Dewey, J. (1916) *Democracy and education*, Macmillan, New York.

Dickinson, E. (2011) "Displaced in nature: The cultural production of (non)place in place based forest conservation pedagogy," *Environmental Communication*, 5(3): 300–19.

Dickinson, E. (2013) "The misdiagnosis: Rethinking nature-deficit disorder," *Environmental Communication*, 7(3): 315–35.

Dobrin, S. and Weisser, C. (2002) *Natural discourse: toward ecocomposition*, State University of New York Press, Albany.

Edge, J. (2002) *Continuing cooperative development*, University of Michigan Press, Ann Arbor.

Grigg, R. (1994) *The Tao of Zen*, Alva Press, New Jersey.

Halliday, M. (2001) "New ways of meaning: The challenge to applied linguistics," in Fill, A. and Mühlhäusler, P., eds., *The ecolingusitics reader*, Continuum, London, pp. 175–202.

James, T. (1995) "Sketch of a moving spirit: Kurt Hahn," in Warren, K., Sakofs, M. and Hunt, J. R., eds., *The theory of experiential education* (3rd edn.), Kendall/Hunt, Dubuque, pp. 75–95.

Milstein, T. (2011) "Nature identification: The power of pointing and naming," *Environmental Communication*, 5(1): 3–24.

Milstein, T. (2012) "Survive, critique, and create: Guiding radical pedagogy and critical public scholarship with the discursive guideposts of ecopedagogy," *Green Theory and Praxis Journal*, 6(1): 3–16.

Milstein, T. and Kroløkke, C. (2012) "Transcorporeal tourism: Whales, fetuses, and the rupturing and reinscribing of cultural constraints," *Environmental Communication*, 6(1): 82–100.

Milstein, T. and Pulos, A. (2015) "Culture jam pedagogy and practice: Relocating culture by staying on one's toes," *Communication, Culture & Critique*, 8(3): 395–413.

Oravec, C. (1982) "The evolutionary sublime and the essay of natural history," *Communication Monographs*, 49, 216–27.

Phillips, A. T. (2014) "Eliding extraction, embracing novelty: The spatio-temporal configuration of natural history," *Environmental Communication*, 8(4): 452–67.

Sarles, H. B. (1977) *Language and human nature*, University of Minnesota Press, Minneapolis.

Smith, T. E. and Knapp, C. E. (2009) *Beyond Dewey & Hahn: Foundations for experiential education*, Raccoon Institute Publications, Lake Geneva.

Stibbe, A. (2010) "Ecolinguistics and globalization," in Coupland, N., ed., *The handbook of language and globalization*, Blackwell, Malden, pp. 406–25.

UPS (1925) *Geyserland: Yellowstone National Park*, Union Pacific System, Omaha.

Vico, G. (1961 [1725]) *The new science of Giambattista Vico*, trans. Bergin, T. and Fisch, M. (3rd edn.), Doubleday, Garden City.

Weisser, C. and Dobrin, S., eds. (2001) *Ecocomposition: Theoretical and Pedagogical Approaches*, State University of New York Press, Albany.

Wong, L. (1986) "Negative–positive dialectic in the Chinese sublime," in Chou, Y., ed., *The Chinese text: Studies in comparative literature*, The Chinese University Press, Hong Kong, pp. 119–58.

7 Further afield

Performance pedagogy, fieldwork, and distance learning in environmental communication courses

Mark Pedelty and Joy Hamilton

Students spend the majority of their time in the classroom sitting and listening. Even in the most highly interactive classrooms, students tend to sit and listen to peers and instructors more often than they move or speak. Typically, online courses create even greater sensory deprivation: students are seated at a desk listening, watching, and typing responses. Therefore, traditional online learning modes are anathema to performance pedagogy, teaching strategies designed to incorporate the students' full range of senses in their learning.

Several environmental communication instructors have taken their students out of the classroom and into field sites. For example, Jessica Prody assigns "place journals" in her Environmental Communication course that incorporate St. Lawrence University's immediate environment. Tema Milstein takes and sends her University of New Mexico students into the field in multiple settings and situations and asks them to acutely observe and interpret their surroundings, combining sensory experience with course readings. Robert Thomas at Loyola University in New Orleans takes students to Bayou Bienvenue Project in the Lower 9th Ward where they apply course concepts to local environmental ecosystems and histories.

Theoretically, digital networking and learning technologies could further liberate students from classrooms to learn in rich contexts more relevant to environmental communication. Why make students spend hours sitting in a crammed classroom when we could be teaching and learning in the places where environmental communication takes place every day? For example, archaeologists prefer to teach students with trowel in hand rather than via PowerPoints about distant field sites. Biologists teach students through field experiments rather than merely assign research reports. Theater instructors help students gain a sense of space and perspective in performance spaces rather than simply lecturing about embodiment and spatial thinking (Varopoulou 2010). As environmental communication instructors, we also desire to have our students learning in places where environmental communication takes place on a daily basis: public parks, superfund sites, boardrooms, public hearings, and so on. In a classroom, such realities can only be simulated. Therefore, we turned our Environmental Communication course into a digitally networked field class, getting students out of the classroom and into rich learning contexts, such as State Parks.

We will refer to student talks throughout this chapter. Two student videos and other materials referred to in this chapter can be viewed at https://sites.google.com/a/umn.edu/4250-moodle-screenshots/. One is an interpretive dance and the other a more traditional interpretive talk; both are shot in three segments at the Stone Arch Bridge over the Mississippi River in Minneapolis. They were chosen to illustrate the wide range of options an interpretive talk format allows.

All information through which individual students could be identified has been removed, including forum posts, assignment submission, and other communication. However, we asked a few students if we could present their videos to readers on the Moodle site. Their videos as well as video and photographic examples from past courses can be viewed in the "Interpretive Talk Resources" module on the Moodle site.[1]

Our goal for the chapter is not to promote the method, however, but rather to work through a difficult question: Is performance-based pedagogy compatible with online learning? While we believe the digital migration from classroom to field has enhanced students' learning overall, it has also made it more difficult to teach performance-based skills that were central to the previous classroom-based versions of the course.

We hope reading this chapter will turn our idiosyncratic challenge into a productive conundrum for readers as well. Many instructors will face the challenge of teaching environmental communication courses online. The centrifugal forces of new technologies and administrators drawn to digital economies of scale—in addition to the genuine benefits of online education—are taking more of our courses off campus. We were drawn by the opportunity new technologies provide for us to take environmental communication education into the field.

Experiential learning and performance pedagogy

The most recent syntheses of the subject—Kolb's (2014) *Experiential learning: Experience as the source of learning and development* and Roberts's (2015) *Experiential education in the college context: What it is, how it works, and why it matters*—explain the epistemological underpinnings of experiential methods and provide empirical evidence that it works. As both Kolb and Roberts explain, the goal of experiential education is to engage the student's entire body and mind in their learning, and to do so in ways that allow them to apply, interpret, and discover course-relevant concepts.

Experiential learning involves placing students in rich learning contexts, those most relevant to course topics, concepts, and skills. For example, based on empirical evidence, Milstein (2005, 233) determined that "sojourners perceive an increase in their levels of communication self-efficacy" after engaging in intercultural encounters abroad. Students, similarly, learn a great deal when they go out and experience the worlds they read about in textbooks and discuss in classrooms.

The premise of our courses is similar: to get students out of the classroom and into new environments, sites where active engagement can increase their communicative competence and knowledge. Drawing on experiential learning projects we have taught over the past fifteen years, we are now using digital networking to turn public land sites into field learning sites where individual students can interact with park rangers, field scientists, park visitors, animals, plants, and ecosystems they might not otherwise encounter within the context of a traditional classroom setting. To borrow Milstein's apt language, students have become environmental "sojourners" rather than fixed bodies in a relatively sterile classroom environment. Using Moodle, social networking and smartphones, these individualized field sites are collectively connected into a networked classroom where students can share their discoveries and knowledge with their classmates.

Which brings us back to performance pedagogy, a method predicated on many of the same assumptions as experiential education, including the claim that students' whole bodies should be engaged in the learning process. Based on past teaching, we have evidence that performance-oriented methods have worked particularly well in our environmental communication courses. For example, quiz scores in a fully field-based, performance-oriented course were, on average, an entire grade level higher than those earned by students in the campus-bound version of the class. Both courses assigned the same readings and the quizzes were also the same. Only the pedagogical strategy differed: one course was field- and performance-based, the other was not. It was not a purposefully planned comparison, but this serendipitous evidence supports what we observed over the years: performance pedagogy in field contexts was more effective than more traditional classroom versions of the same course.

In both the field and classroom versions of the course, we worked with students face to face. That is no longer the case in the digitally networked field course. All of the instruction is online except for one mandatory meeting near the start of the semester and optional office hours visits. Therefore, while the experiential dimension of the course has clearly been enhanced, the performance aspect has become a much greater challenge.

The distance-learning literature provides few solutions to the problem of teaching performance online. Most engaged learning research in distance education contexts focuses on method and medium rather than content and performance. For example, Revere and Kovach's (2011) useful review of "Online technologies for engaged learning: A meaningful synthesis for educators" assumes students use pre-structured learning exercises rather than producing their own media content as well. Students are not required to perform in the courses Revere and Kovach examine. Given the goals of our environmental communication course, it is very important for students to produce original media projects that involve presentation in front of the camera. Rather than evaluate others' interpretive talks as the endpoint, we want students to research, script, perform, and record their own interpretive talks at their field sites. Analyzing existing media is useful, and we do ask students to assess professional ranger talks, but the act of producing an original interpretive talk greatly aids their sense of discovery, skills building, and conceptual learning.

By combining smartphones, a robust electronic learning platform, and field learning, we hoped to "include elements of performance and theatre" in computer-mediated education, as Rod Sims (1999, 257) proposed in the early days of distance education. However, the "performance pedagogy" literature has not intersected significantly with the distance learning literature. MacDonald's (2014) "Performance as critical posthuman pedagogy" comes closest, but even that work is more about critiquing technology than deeply engaging with it. Conversely, in our online course we seek to use digital technologies to allow students to communicate critically about environmental issues from distant, biodiverse field settings.

As instructors, we know the best way to determine whether we understand a concept is by assessing how well we explain it to our students and colleagues. The same goes for student learning and comprehension. In most cases, it was apparent that students who did well on the interpretive talk assignment also learned key course concepts. Skills acquisition and conceptual learning are mutually constitutive.

Theoretically, performance pedagogy could become an effective complement to experiential education. Our working definition of experiential learning has involved various combinations of discovery learning and public performance. Assignments have ranged from producing environmental benefit concerts to ranger-style interpretive talks, but in all cases students have produced and performed original content rather than fulfilling a pre-set exercise or digital storytelling, a form which usually implies aggregation of existing materials rather than independent production and performance.

Our goal is to help students develop into fully functioning environmental communicators, people who competently explain scientific research, an institution's environmental goals, their neighborhood's concerns over pollution, and so on. In other words, we want students' learning to mimic "real world" conditions they are likely to face in their future professional, political, and social lives. Rarely are environmental issues communicated through something akin to digital storytelling or pre-set game-based exercises. The art of human communication remains key. Therefore, we are concerned with helping students develop the basic skills required of professional environmental communicators. We find that students are rather adept at digital storytelling and social media by the time they enroll in our course but often lack the public speaking and media production skills expected of environmental communicators.

We do not demand production expertise, per se; a smartphone video recording is just fine. However, we expect students to develop the sorts of on-camera performance that approximate the daily work of professional environmental communicators. We believe most students achieved this in our past face-to-face courses, and we wanted that to happen within the online course as well.

However, more students fell short of that goal in the online course than in the classroom versions, despite commensurate or greater achievement in other aspects of the course. In person, performance inhibition can often be overcome with concerted effort. Over the course of a term, students get incremental help and feedback to build solid performance skills. When working face to face, subtle cues and structured performance exercises allow students to become effective environmental communicators via deep interaction with instructors and peers. We found it difficult to replicate that level of interaction online.

It was a tall order, therefore, to ask students to engage in fieldwork, conduct library research, write about the results of their research, script out a talk, rehearse it, and then film the results. We have helped students to achieve that entire sweep of activities in the controlled confines of the studio, classroom, and campus, but now our students are physically distant and have the element of field study to deal with as well.

An evolving sense of performance

The best way to explain what we mean by "performance pedagogy" is to briefly describe the evolution of COMM 4250/5250, a course in environmental communication. Much of the hybrid undergraduate/graduate course developed in Identity, Community, and Culture: Connections in the Arts and Humanities, a course Pedelty taught three times starting in 2000. Environmental identities were foregrounded in the course, emphasizing connections to watershed communities. Students prepared a wide array of solo projects, in the end performing monologues, dances, songs, and poems on the Washington Avenue Foot Bridge above the Mississippi River Gorge. The course was co-taught with theater director and speech instructor, Heather Dorsey.

After teaching Identity, Community, and Culture: Connections in the Arts and Humanities, Pedelty began teaching Environmental Communication courses, including a graduate seminar, two seminars for first year undergraduates, and three senior-level courses. In these Environmental Communication courses, "performance" took on an additional meaning: professional application. Skills development and professional efficacy were emphasized in addition to performance in the more theatrical sense. Although each semester a handful of students still choose to perform music and/or create visual art as part of their term project, most opted for more professionally relevant modes of delivery.

In each of the first year seminars, students designed, developed, rehearsed, and performed "interpretive talks" on topics of their choosing. As explained above, that assignment has become the center of our new digital field course. In addition to being a central focus of the first year seminars, the interpretive talk assignment was taught as a unit in an Honors Public Speaking course. The assignment allowed students to speak in a more natural social and physical setting to start out the semester. Students enjoyed speaking within contexts connected to their subjects of interest, such as a talk about the history of brick presented next to a brick building, a discussion of bicycle commuting around a bike rack, and a speech about urban runoff in front of a raingarden.

A modified version of the interpretive talk assignment was integrated into a graduate seminar in environmental communication as well. There is a rich tradition to draw upon when integrating performance-based pedagogy into graduate seminars. For example, New York University's Richard Schechner and Victor Turner turned graduate Anthropology and Performance Studies seminars into performance workshops during the 1960s and 1970s (Harding and Rosenthal 2006, 327). Some of the most successful grad level projects at the University of Minnesota have professionally relevant topics, such as a bio-engineer arguing the merits of sustainable supply-chains to executives or a mining manager explaining permitting processes to the public.

However, there is a significant gap in our use of performance-based methods. We have not used performance extensively in upper division environmental communication courses. The first upper division course required students to complete worksite ethnographies with the Resilient Communities Project (www.cura.umn.edu/RCP) and City of Minnetonka in a service-learning context, which required us to drop the creative performance component for sake of time and focus. Students monitored and observed environmental communication but did not engage in communicative development more directly beyond ad hoc discussions with clients.

The following year, we collaborated with the local environmental nonprofit Metro Blooms (now Blue Thumb). Students designed and installed a large rain garden on campus as part of the University of Minnesota "Living Laboratory" program. The garden became a focus for learning about ecological issues as well as a way to practice environmental communication skills on interpersonal, community, and institutional level. However, the last classroom-based iteration of COMM 4250/5260 required students to study an issue in depth and communicate it to the community using whichever performance mode they preferred, from traditional tabling to original songs, all performed on the University Mall.

Further afield: the ups and downs of digital distance

Some 29 percent of the environmental communication courses Hamilton surveyed require students to complete projects outside of the classroom (see Chapter 3). Assignments include reflection journals, having students unplug from technology, service learning with environmental organizations, and field trips to public lands. Various iterations of our environmental communication courses have involved fieldwork as well. For example, Pedelty took a group of students to London and Cornwall, England, where they researched local topics and performed interpretive talks at various field sites, such as a talk about e-waste presented in front of the giant WEEE (Waste Electrical and Electronic Equipment) man sculpture at the Eden Project (2015) to another about composting, performed in front of the world's largest composting pile, in Kew Gardens, London. Hamilton led a summer service-learning course about hydraulic fracturing, coal mining, water rights, and agriculture in the U.S. Southwest for the University of Colorado's Community Leadership Program. Students met with miners, environmental activists, local politicians, journalists and NGOs to gain a comprehensive understanding of the region's complex environmental challenges. These field conversations provided a context and experience that exceed the limitations of a classroom setting. Although these field seminars were wonderful teaching and learning opportunities, they could never be scaled up for large groups of students. That is another reason why we instituted the digitally networked field course. We believe that online instruction can effectively bring field-based learning to more students. Unfortunately, as we have redesigned and taught the course, we have also struggled to retain performance-oriented environmental communication learning methods that were key to the course's past success.

We are seeking to overcome some of these challenges with a simple, somewhat individualized approach: each student finds his or her own local field site and studies a site-related environmental issue. "The field" in this case is any public land or waterway in Minnesota the student chooses; this also fulfills the University's Land Grant mission. Students interact with each other virtually while also learning from experts and ecosystems. Much was gained by turning the outside world into a digitally networked classroom. Many aspects of the classroom-based courses seem to be transferring well.

We required students to research an issue on site and use academic literature. They are required to write weekly reports that step them through the research process systematically. As part of the same weekly reporting system, students write outlines and scripts for their interpretive talks, rehearse the talk, and record it on video. As park rangers demonstrate, the interpretive talk format can adapt to incorporate many types of public performance, from detailed scientific explanation and historical re-enactment to artistic interpretation and musical performance. Students are allowed to choose virtually any environmental issue that interests them and relates in some way to their field site.

Students are required to introduce and explain a site and related issue to the public, much like a museum docent or parker ranger would do. Those so inclined bring in poetry, music, or theatrical re-enactment, although fewer than we had hoped. Others go more deeply into the environmental science underlying a given issue. In other words, the interpretive talk is adaptable enough to suit most students' interests, needs, and majors, from the arts to STEM (science, technology, engineering, and math). Meanwhile, the course presents a systematic survey of the inter-discipline of environmental communication, including readings, video lectures by the instructor, video interviews with guest experts, and a weekly podcast to keep the course up to date. Working with a team, Pedelty produced each of the videos before the semester began. The team included Sara Schoen and Paul Ching working under the auspices of Experiments in Learning Innovation (ELI), a program developed by our university

Provost, Karen Hanson, to provide support for online course innovations. Working under the auspices of an assistantship, Hamilton developed much of the Moodle infrastructure and supplementary support for the course. In addition to their site-based projects, we asked students to incorporate chapter topics from Robert Cox and Phaedra Pezzullo's (2015) *Environmental communication in the public sphere* into their weekly reports. We also tested their knowledge of the reading and videos via weekly multiple-choice quizzes.

Pedelty met with each student to start the semester, which was essential in terms of initiating personal contact. We produced model interpretive talks and presented them in the second week of class. Those model talks, additional video examples, and a weekly audio podcast were designed to demystify the assignment with a "making of" dimension to help students create their own interpretive talks. Meanwhile, professional interpretive talks culled from the web and others captured locally provided aspirational examples. Interviews with environmental scientists, rangers, artists, performers, and colleagues in the ecocritical humanities provided additional ideas and advice for students. These and several other performance-oriented supports were made available on Moodle and directly built into the course.

Did these online supports make-up for the loss of face-to-face workshops? Based on a formal course assessment there is some reason for optimism. The more a student engaged in the course, the more likely he or she was to produce a successful interpretive talk. We measured each student's total Moodle activity, which is the most accurate way to determine their individual level of engagement with course resources. This measure accounts for all course activities and is independent (in terms of grade) from the interpretive talk grade. In other words, total Moodle activity (V_1) and interpretive talk grade (V_2) are not duplicative measures. None of the weekly quizzes, weekly report grades, working group grades, forum activities, reading, watching, listening, or game activities were counted within the interpretive talk grade; only the actual video recording of their performance was assessed. Therefore, Moodle-based engagement is a useful indicator of the degree to which students took advantage of course supports and our best indication of whether or not using those materials and methods helped them produce a quality talk. That was indeed the case. A Fishers Transformation Test found a positive correlation between individual students' Moodle activity (V_1) and interpretive talk scores (V_2) ($p = 0.021$).

We are not claiming that course supports are the only factors that led to quality talks. As is usually the case, it is unclear whether the best students—as measured by level of course engagement—also happened to do well on the interpretive talks or if it was their more intense use of course supports that actually produced that success. The most that we can say is that the two variables are positively correlated.

Within the data are several outliers, of course. For example, there are several students who barely looked at optional supports yet did very well in their final performance. Those students may have come into the course with adequate skills and knowledge to complete the interpretive talk assignment without a great deal of support. However, several of those students failed their rough cut before going on to succeed based on instructor and peer feedback, often after misunderstanding the parameters of the assignment and, generally speaking, the genre. Having ignored the systematic supports for the assignment, several of the rough cuts looked nothing like an interpretive talk. However, with direct feedback after the rough cut many of those same students did very well, in some cases excelling.

Therefore, even in those outlying cases, it would have benefitted the students to look at optional resources and doing so would have helped us to teach them more effectively. We could have spent less time explaining to those students what they could and should have known by the time they reached the rough-cut stage. They could have put their time and energy into advancing their research and presentations.

Other outliers included students for whom English is a second language. Second-language students used course supports to a very significant degree yet in several cases did not perform as well on the interpretive talk assignment. While first language is taken into consideration in the grading, language facility inevitably impacts the student's ability to prepare and perform a live presentation in English. It is worth noting that the most exceptional performers among second-language students came to the instructors' office hours often to receive direct help. Either the classroom version of the course or a supplementary workshop would greatly benefit these and other students who require additional face-to-face support. This is somewhat of a problem because second-language students appear to opt for online courses more often than their peers.

With those caveats in mind, we draw the conclusion that course supports were a fairly useful replacement for face-to-face performance support methods. However, qualitative assessment of the videos indicates uneven results. It appears that the online supports could not replace face-to-face performance instruction for several students. That assessment rests on comparative assessment of the final interpretive talks, a gestalt approach rather than rigorous assessment (it would be prohibitively time consuming and expensive to arrange an external, blind, and comparative review of samples of interpretive talks culled from our online and offline courses). Nevertheless, we believe, based on teaching in both contexts, that many of the lower performing students in our online course would have been much more successful in the assignment if they had experienced face-to-face performance workshops.

We had hoped that the opportunity to use a smartphone or camera on site for multiple takes and retakes might put more control over the performance into the individual student's hands. Learning to perform is an iterative process. Even in the classroom, students gain performance skills as much through repetition as through feedback from peers and instructors. With smartphone in hand and on site, students could work on the final performance until they reached a high level of proficiency, greater proficiency than a one-take public performance allows. In other words, we hoped that putting that control into student's hands would produce *better* talks than in the classroom context. However, the ability for students to have that iterative control appears less important than the performance training made possible in face-to-face instruction. Peer and instructor feedback on the "rough cut," the first video submission, helped a great deal, but many students in the online course struggled with their final presentations to a degree that students in past performance-based versions of the course simply did not. In moving the course online, we lost the "co-present" qualities of the classroom that allowed us to slowly coax students' inner performer out of their shell. In the past, it took many weeks of careful face-to-face work with students to help them overcome their inhibitions and gain the confidence to perform while being watched. In that regard, at present there might be no effective replacement for that in the online environment.

As for other measures of course quality, student evaluations were in the same range as past, offline offerings. On a scale of 1–6, with 5 being "agree" and 6 being "strongly agree," course medians were 6 for all categories and means ranged from 5.37 to 5.76. Common complaints in the qualitative feedback related to workload, a problem we deal with below. "This isn't a complaint," explained one student, "but there is A LOT of writing and work each week for this course. I think it should be 4 credits." The student had a valid point. As for comments related directly to the project and the main question in this chapter, several students found the combination of broad parameters and high demands problematic in the interpretive talk project. Once again, we agree.

However, the overwhelming majority of comments were very positive, with students noting that feedback on their interpretive talk was helpful and essential. Positive comments about individualized feedback indicate that students value dyadic relationships even in, or perhaps especially in, the online context.

Yet, if we cannot help students develop effective final presentations—our best measure of whether or not they have become effective environmental communicators—then the pedagogical premise of the entire course comes into question. It is difficult enough to give an interpretive talk to a group of classmates one has come to know and trust over many weeks in the classroom, or to perform an environmentally themed song in front of the student union, having received encouragement and help in the creative incubator of the classroom all semester. It is risky enough to stand atop a seawall in Cornwall to talk about rare earth metals, having had several weeks to test run the performance with instructors and classmates. However, it is quite another thing to film a presentation on one's own, knowing full well that it will later be watched. Perhaps we are asking too much, but the students' employers will soon demand effective communication skills as well. Therefore, we need to think seriously about whether or not to keep the entire interpretive talk intact or modify it to fit the online format and single semester schedule.

Solutions

The course has evolved in very positive ways, bringing us to this mostly positive point in curricular development. Students have been liberated from the classroom, digitally assisted to do important research and writing, to explore and learn in rich field contexts. Yet, our performance conundrum remains.

We have therefore used the results of this assessment to revise the interpretive talk assignment for spring 2017. One conclusion is that the semester is too short to rigorously research a site and issue, write about it, outline and script a talk, rehearse the talk, revise it, film it, revise it again, and film a final version. It is partly a matter of medium. Video presents certain additional difficulties that might be getting in the way of other learning activities and is less amenable to staged development over the course of the semester.

The interpretive talk assignment will consequently be converted into an audio production. That way it can be developed on a weekly basis, allowing us to provide more continual feedback while minimizing performance, media, and tech variables that are difficult to address effectively in addition to all else that is required in the assignment and course. Just as we were able to use weekly audio podcasts to keep the course up to date and dynamic, converting the interpretive talks into audio projects will allow students to spend less time on visual performance and tech concerns and more of their very limited end-semester time on crafting an audio-only presentation.

While there is a great deal of popular rhetoric concerning the ease of consumer video technologies and social media platforms, the course served as a reminder that we are still not at the point where students bring sufficient visual literacy and video production skills with them into a general studies course. From performance to upload, video capture remains a more technologically involved and time-consuming process than does audio production. Journalists have long known this, finding video more cumbersome and limiting than audio when it comes to field reporting (Pedelty 1995). It is a lesson we relearned in this course.

We experienced the difference between video and audio ourselves while producing video lectures and weekly podcasts. Audio was more adaptive, accessible, individualized, and dynamic compared to the videos, and allowed students to listen while driving, cooking, or exercising. They were also far easier to produce than the videos, allowing for time to be put into informational content as opposed to visual performance. The same is true for students. We will simply have to cut visual performance training from the course learning objectives.

In the next offering of the course, therefore, students from early on will start developing, practicing, submitting, and receiving feedback on audio-based interpretive talks. Discussions are underway to work with the State Park system to have students produce site-specific audio presentations at their field sites. Park visitors will be able to stream the audio on site. While losing the visual dimension is a concession to time and technology, it is also an opportunity to focus on sound and voice in ways that could benefit students. Already within the course there is a discussion of soundscape and sound studies. We can now integrate that area of study more fully into the interpretive talk assignment. By targeting a vocal performance in the audio recordings, we can help students learn to capture sounds and reference soundscapes in their audio presentations.

We also need to find more ways to integrate optional supports directly into the class and make them mandatory for all. Making supports optional signaled to many students that they were non-essential. As indicated by the assessment data, they are instead essential to success in the interpretive talk project.

The same goes in regard to narrowing the parameters for the assignment, including site choice. Too many students chose conveniently located parks instead of seeking out options with sufficient interpretive resources. The same students who would benefit most from the presence of a visitor's center, rangers, placards, websites, nature trails, and public events were the least likely to choose a site featuring those interpretive resources. Many students ignored advice presented in the syllabus, early podcasts, and the introductory video. Conversely, more motivated students tended to choose State, National, and regional parks that offered those helpful resources, gaining another leg up on the assignment. In 2017, therefore, a list of parks will be provided.

As for what worked and will remain in the course, the list is too long to present here. Replacing the classroom with digitally networked field sites worked very well for most aspects of the course. Even areas of the course we feared might not remain sustainable year to year due to their demands on instructors' time and effort turned out to be more feasible than we imagined. For example, it took a great deal of time to meet with over 100 students individually during the third week of the course, but it was an invaluable opportunity to establish meaningful individual contact and help students tailor their projects. Also, watching each student video twice (Pedelty), first during the rough-cut period and then the final version, was well worth the extensive time it required (and more entertaining than reading and grading long term papers).

We do require fifteen to twenty pages of writing, but it is completed and graded less painfully than the all-at-once term paper, one page per week. Having three teaching assistants provide weekly feedback on those research reports was essential and invaluable; the course simply would not have been possible without Christian Angelich, Diane Cormany, and Joy Hamilton providing feedback to about thirty students each per week.

Nevertheless, that scale of operation might be problematic, as predicted by our department Chair, Ron Greene, when we were in the planning stages. We might therefore have to reduce enrollment in the next offering back to forty students. It is a challenging course that requires weekly work and assessment. Each student logs on to the course Moodle site an average of 1,212 times over the course of the semester. In other words, the average student was engaging in no less than seventy-five Moodle actions per week, including total views, posts, and other activities. They log on to watch videos, listen to podcasts, take quizzes, read supplementary articles, post forum comments, submit weekly reports, take part in educational gaming (a "Public Participation Game"), list a make-up activity, and so on, a level of activity that led to high attrition early on as less motivated students fell behind in their work. Accustomed to courses with high stakes midterms and many weeks without graded assignments, students who did not do their required weekly work from the start were in danger of failing the course within the first six weeks, taking make-up options into consideration. Of the 125 students who initially enrolled, only 98 finished the course.

However, the remaining students achieved an 84 percent score on average, solidly in the B range. These were the students who completed the interpretive talk assignment. While it was useful to have a mechanism to identify non-participating students early on, clearly our requirements were a bit high given the number of activities students needed to engage in to succeed in the course as a whole. The "capstone course experience" might have instead become an overly difficult right of passage for graduating seniors who make-up the majority of the course enrollment. Perhaps we were trying to create the illusory "perfect course" (Miller 1993), overloading a single class that should instead be broken into two semesters: a larger general studies course followed by an advanced field course.

Conclusion

Our struggle to balance an overly large set of learning objectives within a single course has helped us to understand better what we teach and why we teach it. The challenge of taking the environmental communication course online and into the field has forced us to critically examine what is actually being learned. It has required us to let go of aspects of the course we considered fundamental, including visual and kinesthetic approaches to performance. Letting go of that element should allow the experiential, field-based dimension of the course to achieve its full potential.

Despite the performance conundrum explored in this chapter, we are very happy that we made the transition to a digitally networked field context and we recommend it to colleagues. Students were introduced to environmental communication in thematically rich field contexts, were able to pursue issues they care about, benefitted from each other's field-based learning, and gained a broad introduction to environmental communication. Although this chapter focused on the challenge of teaching performance skills online, in almost all other aspects the transition to a digitally networked field format seems to have advanced student learning.

Note

1 We would like to thank the following students for giving us permission to put their videos on the Moodle site for this chapter. They presented outstanding interpretive talks: Maxwell Burrows, Zachary Carlstrom, Kristin Hussong, Joshua Kelley, Natalie Lande, Lauren Lee, Josie Milan, Bay Miskowiec, Allie Noetzelman, Sejla Sacic, Trevor Squirer, Ashley Strohmayer, Abby Taylor, Mackenzie Trammell.

References

Cox, R. and Pezzullo, P., eds., (2015) *Environmental communication and the public sphere* (4th edn.), Sage Publications, Thousand Oaks.

Eden Project (2015) Giant Sculpture Made of Waste. Available at: www.edenproject.com/visit/whats-here/giant-sculpture-made-of-waste (accessed July 24, 2015).

Harding, J. M. and Rosenthal, C. (2006) *Restaging the sixties: radical theaters and their legacies*, University of Michigan Press, Ann Arbor.

Kolb, D. A. (2014) *Experiential learning: Experience as the source of learning and development*, Pearson Education, New York.

MacDonald, S. M. (2014) "Performance as critical posthuman pedagogy," *Text and Performance Quarterly*, 34(2): 164–81.

Miller, J. E. (1993) "Tradeoffs in student satisfaction: Is the "perfect" course an illusion?," *Journal on Excellence in College Teaching*, 4: 27–47.

Milstein, T. (2005) "Transformation abroad: Sojourning and the perceived enhancement of self-efficacy," *International Journal of Intercultural Relations*, 29(2): 217–38.

Pedelty, M. (1995) *War stories: the culture of foreign correspondents*, Routledge, New York.

Revere, L. and Kovach, J. V. (2011) "Online technologies for engaged learning: A meaningful synthesis for educators," *Quarterly Review of Distance Learning*, 12(2): 113–24.

Roberts, J. W. (2015) *Experiential education in the college context: What it is, how it works, and why it matters*, Routledge, New York.

Sims, R. (1999) "Interactivity on stage: Strategies for learner–designer communication," *Australasian Journal of Educational Technology*, 15(3): 257–72.

Varopoulou, H. (2010) "Landscape with students and teachers: Landscape, environment and performance pedagogy in a Greek Summer Academy," *Performance Research*, 15(4): 123–30.

8 Arts-based research in the pedagogy of environmental communication

Geo Takach

A nascent but natural alliance is arising among the study, practice, and teaching of environmental communication and arts-based forms of engagement. These forms offer enormously effective potential to provide highly expressive, nuanced representations of the world and life on it; to cast fresh eyes on entrenched perspectives, inviting questions and conversations; and to engage, inspire, and activate audiences, both within the academy and beyond, towards essential environmental change. Accordingly, this chapter makes the case for arts-based research in the pedagogy of environmental communication by outlining how environment and art are linked in scholarship, practice, and pedagogy; highlighting some key goals, benefits, and challenges of arts-based research and pedagogy; and providing examples of, first, one of the many forms that such work can take (drawing on my short, eco-documentary film, as I am also a professional filmmaker)[1] and, second, how arts-based research might be integrated into pedagogy in the field (drawing on a capstone assignment developed for my graduate course in environmental communication).

Linking environment, art, research, and pedagogy

So why can the arts be particularly fruitful in teaching environmental communication? It starts with the strong links among environment, art, research, and pedagogy, all of which, of course, involve communication. In describing the profound connection between our natural or built environments and artistic expressions of how we experience them the philosopher Edward Casey (2004) suggests that those expressions can help us find better ways to live on Earth and concludes that in this sense all art can be seen as environmental. Given that the core purpose of research is "to create knowledge that matters," we can conceptualize art as "a creative and critical 'research' practice that opens up new pedagogical possibilities" (Sullivan 2012, 147). Thus, as art can help us to envision and bring about a more just world (Quinn et al. 2012; Dewhurst 2014), cultivating our students' creativity through arts-based pedagogy can ultimately incite transformative social change (Sullivan 2010). This would certainly include acting on environmental concerns.

Art and environmental issues are linked further to education and activism. Reflecting its roots in educational research (Eisner 1997), *arts-based research* has been connected to environmental education (Inwood 2010), and 'green art' has been viewed as an expression of eco-justice in a pedagogical context (Finley 2011a, 2011b, 2012). Integrating artistic practice into education and advocacy on ecological issues addresses McKibben's (2005) call for artistic expressions of climate change—which we can extend to environmental issues generally—to boost understanding and feeling, and thereby create the kind of political change that art helped to inspire in the case of AIDS over the last two decades. He urges us to bring environmental issues to the public on a broader, more emotional level, to "register in our gut" and become "part of our culture." Again, the aims of green art are allied to those of green scholarship. Moser and Dilling (2007) push critical scholarship toward activism in calling for communication scholars to help engender a collective will to reduce greenhouse-gas emissions, not only by reframing the way that the media take up the discourse of climate disruption (Boykoff 2007), but also by showing ways to structure communication that promote agency, and thus inspire widespread public action toward ecologically sustainable policies and practices (Foust and Murphy 2009). Finley (2011a, 2012) advocates a pedagogical "ecoaesthetic," which draws on artistic practices to open more spaces for public dialogue, debate, and action on ecological concerns.

Foundations of arts-based research and pedagogy

Premised on the idea that "many of the greatest contributions to human understanding have been generated by the arts" (McNiff 2008, 38), arts-based research is defined as:

> the systematic use of the artistic process, the actual making of artistic expressions in all of the different forms of the arts, as a primary way of understanding and examining experience by both researchers and the people that they involve in their studies.
>
> (Ibid., 29)

This hybrid methodology transcends art and science (Eisner 2006) and in a sense counters the separation of artistic and scientific inquiry that occurred during the (so-called) Enlightenment period of Western history (Barone and Eisner 2012). This research paradigm arose because traditional academic vocabulary "fell short in its ability to capture and communicate the complexity of human experience in all its diversity," making little impact on participants or their communities (Knowles and Cole 2008, 57). That vocabulary tends studiously to avoid emotion, a core of the human condition and thus a vital avenue for human engagement.

Drawing on a richer palette of "creative intelligence and communications" lets arts-based methods and teaching "generate important information that often feels more accurate, original, and intelligent than more conventional descriptions" (McNiff 2008, 30). Eisner (1997, 8) suggests five advantages of arts-based inquiry, which we can extend to arts-based pedagogy:

- an alternate form of presentation to shape experience and expand understanding;
- a deeper sense of detail;
- "productive ambiguity";
- a wider variety of questions to ask and explore; and
- exploitation of individual aptitudes beyond the literal or quantitative.

To these, Brandt (2006) adds that arts-based research can inspire creativity, recover repressed histories, build community and strengthen social movements, while serving in itself as a holistic form of action. Of course, arts-based research and teaching also come with challenges, such as underdeveloped theoretical foundations (Leavy 2015), issues around the level of artistic skill and expertise required (Woo 2008), criteria to assess the contributions of arts-based research (Lafrenière and Cox 2013), logistics of the scholarly publication of results, and potential resistance to less orthodox forms of presenting research (Eisner 1997).

The door to arts-based research and pedagogy in environmental communication is wide-open. With its focus on human relationships to nature and other spaces, environmental communication has already studied constructions of the environment in diverse artistic efforts such as painting (Hodgins and Thompson 2011), music (Pedelty 2012) and literature (Guignard and Murphy 2009). These links are bolstered by critical approaches common in both arts-based research—which is counter-hegemonic to positivistic, scientific research, as well as socially activist (Finley 2011a, 2011b, 2012)—and environmental communication research—which challenges anthropocentric assumptions frequently underlying dominant, popular representations of the environment (Corbett 2006).

Environmental arts-based research

So what can arts-based research look like in practice, and how can it be applied to engage our students? It can take on an infinite number of disciplinary and pedagogical forms, but one example that stems from a qualitative-research project is my eco-documentary film interrogating the tar sands in Alberta, Canada. Called the largest industrial project on Earth (Leahy 2006), the sands comprise the biggest known deposit of synthetic oil—a tarry substance called bitumen—the logistically and technologically complex extraction and refining of which provides a massive financial bonanza for industry, investors and government coffers (Canadian Energy Research Institute 2014), but raises grave concerns over the devastating environmental impacts (e.g., Avery 2013). In a society that communicates voraciously through images, the highly polarized and politically charged debate over mining and pipelining the tar sands[2] is manifested in critical, independent documentary films attacking Albertans' stewardship of the resource, countered by online advocacy videos from the Alberta government and the oil industry defending that stewardship (e.g., Radford 2008; Canadian Association of Petroleum Producers 2010; Takach 2014). My research project included in-depth interviews with the directors and producers of these films and videos, along with a visual and a framing analysis of their work. Hearing the diverse, conflicting voices in my head and dreaming of reconciling them, I resolved to bring them to life for an audience beyond the academy—and to have them (and hopefully the rest of us) engage in a dialogue aimed at depolarizing the rancorous public debate and, ultimately, achieving more responsible environmental policy around fossil-fuel extraction. The resulting short film, *Voices from the Visual Volley: Filmmakers, Tar Sands and Public Health* (Takach 2013) explores that debate through the interplay of image, voice, text and music in an experimental collage, while responding to the literature's call to position environmental concerns as public-health issues and to engage a broader public in productive dialogue and action on more sustainable environmental practices (Cardwell and Elliott 2013).

Abstracting the voices of those filmmakers and public-relations professionals from their interviews and transposing them to a disembodied, shifting landscape and soundscape lets me project those visual storytellers' perspectives onto a canvas differing somewhat from their chosen media of film and video. Eight drama students from two local universities, who were encouraged to bring their own emotions and reactions to voicing the diverse environmental messages, voiced my interviewees' conflicting words and values. Thus, the students' performances included, for example, defiance, concern, embarrassment, despair and shame in portraying the dissenting environmental filmmakers, and pride, awe, bombast, dismissiveness and irony in portraying representatives of the oil industry and the Alberta government. The voices were presented over a collage of stylized text capturing their words and presented alternately on opposite sides of the screen, as in a volley in tennis (Figure 8.1).

Created by a recently graduated visual-communication designer, the visual backdrop was an impressionistic, enhanced photographic image of a landscape (representing the boreal forest being gutted by surface-mining the tar sands) that becomes progressively degraded as the dialogue (and postponed remedial environmental action) continues. The aural backdrop, composed by a teaching colleague, was a chilling, synthesized musical soundscape capturing, in turn, the escalating tragedy of maintaining the status quo, a growing willingness of both sides of the volley to engage in meaningful discussions (accompanied visually by a brightening of the skies before the sun sets over the gutted landscape), and concluding that the greatest danger is silence. A professional video editor and editing instructor helped to assemble the film. It premiered at an international exhibition called InSight2: Engaging the Health Humanities, at the University of Alberta in 2013, and was screened later at scholarly conferences on environmental communication and film studies, university guest-lectures, and another art exhibition. At the first exhibition, gallery-goers were invited to record their

Figure 8.1 Screen shot from *Voices from the Visual Volley*.

Source: photography and design: Iwona Faferek, used with permission.

own vocal contributions to the conversation on a vintage cassette recorder placed next to the screen.

A pedagogical example

From a project created with students outside of the classroom, we turn to an example of how arts-based research can be incorporated into a course on environmental communication. Here I draw on a capstone assignment created for a pilot course developed for a graduate program in communication and technology at the University of Alberta. The assignment requires the student to produce an original work of research-creation, presentable in any form or medium, and based on two key foundations:

- The work must include an inquiry into some aspect of environmental communication covered in the course, be it a general topic like the theoretical foundations of the field, a more specific one like 'green' advertising, or any related area.
- The work must draw on (but not necessarily follow slavishly) an understanding of essential artistic conventions of the form or medium of the student's choice, for a dramatic through-line in narrative fiction.

In setting parameters for students' choices around those two key foundations, instructors should stress the requisite symbiotic connection between the content of the inquiry to be undertaken and the form in which its results will be expressed. For example, a study of news-media representations of public attitudes toward rising sea levels in Florida may yield stories and ideas that lend themselves to:

- reflection in a personal essay by the student;
- dramatization in a one-act play featuring reporters commenting on their coverage;
- a sequence of songs, each capturing different perspectives on the issue;
- a series of sketches or paintings depicting those perspectives; and/or composites of people espousing them; and scores of other formats.

No one medium is necessarily better than another; each case will depend on the sensibilities, affinities and talents of each student. The circumstances of each inquiry may suggest that form follow content or vice-versa, but, in the end, the two should be congruent. For example, a study of people's reactions to images of landscapes wracked by hydraulic fracturing (fracking) could lend itself to presentation in a visual medium, in which those images can be seen, and/or an aural medium in which participants' audible expressions can be heard.

For the capstone assignment in my course, students must submit a project in two parts. The first part is the aforementioned research-creation, capturing the overall thrust, or at least one or more key aspects of their research in an arts-based format. This may be situated in any medium, but, in all cases, a physical or electronic copy of the underlying work must be submitted. (If the work is a script for a performable work such as a play, a film or a dance, then students should submit that script.) The second part is a critical reflection on students' learning in the course, which has four requirements:

- They must provide a rationale for their choice of topic and arts-based format for the first part of this assignment;
- They must discuss how that format helped or hindered the achievement of their communicative goals for their assignment;
- They must share how their thinking on environmental communication has evolved since starting the course;
- Finally, they may include any further reflections that they deem relevant.

Students present a brief extract of their assignment to the class in whatever format they choose, be it reading from their script, dramatizing part of it, showing a video clip of their work, etc.

Challenges around evaluating such an assignment are hardly new, as any fine-arts teacher can relate. Although interpreting art is highly subjective and involves varied and complex considerations, when it comes to assessing the effectiveness of a work of environmental communication in light of specific course material, instructors can apply the same kinds of standard or rubric as in marking assignments such as essays, personal reflections or communication plans. This can involve traditional grading criteria such as the scholarly rigor of the student's application of lessons learned from the course material, the appropriateness of both the content and form of the communication to the task at hand, or the general professionalism of the work's presentation. Finley (2011b) reminds us that ultimately the value of critical arts-based inquiry is its utility to the community served by the research. As an example of how we might assess arts-based research by our students, in addition to the foregoing factors, my rubric for this assignment draws on criteria discussed by Lafrenière and Cox (2013). These criteria are based on three kinds of considerations:

- First, normatively, we can ask, "Does the artistic piece derive from data collected, interpreted and analyzed through rigorous and ethical qualitative or other research practices?" (ibid., 6).
- Second, substantively, "Is the research work created and produced according to the technical and artistic properties of its genre(s)?"
- Finally, from a performative perspective, "Does the artistic work have an effect on the audience that enhances appreciation for the experiences of research participants and/or the overall study's findings?"

The weighting of these and the traditional grading criteria are in the instructor's discretion, but in applying the second (substantive) criterion, we should remember that, in most cases, our students will have had limited time to learn the finer "technical and artistic properties" of their chosen form. While it is certainly desirable for students' work to reflect fundamentals of that form (as in my example of a dramatic through-line in narrative fiction), I suggest that this is one factor in which their effort may merit more points than their execution.

In allocating grading weight (and, consequently, students' time) to this assignment relative to the entire course, instructors should consider whether to require students to conduct a fresh inquiry or to base the project on research such as data, stories and perspectives that they gathered for an earlier assignment in the course. The latter option becomes attractive if course time for arts-based work is more limited. One must also consider that students' varying (and, in some cases, negligible) degrees of acquaintance with diverse artistic disciplines may require them to take additional time to (re-)familiarize themselves with the foundational conventions of the one of their choice. In any case, it is important to note that arts-based research is considered a valid form of inquiry in its own right, rather than merely a means of expressing researchers' findings (Haseman 2006). The artistic creation and expression becomes a critical part of the research! The artistic form allows and actually encourages representations that transcend research representations denuded "of emotion, of sensuality, of physicality" (Knowles and Cole 2008, 57), while also evoking the more visceral, emotional reactions among audiences demanded in Bill McKibben's (2005) appeal for works of art that "register in our gut" and become "part of our culture."

Unleashing environmental inspiration

From its roots in pedagogical studies, arts-based research stands to help us to communicate environmental information and concerns more engagingly and viscerally to audiences beyond just the classroom. Building on established associations between art and the environment, teachers of environmental communication can integrate arts-based research into students' learning, not only to further curricular goals, but also to encourage students to share their work on a broader stage after the course, through a kaleidoscope of artistic forms of expression, aimed at improving the environment and life on Earth.

In closing, I invite instructors and scholars to consider some questions:

- Where can you see arts-based research methods being useful in teaching environmental communication?
- How might such methods help to inform and share your own scholarship and/or practice?
- Finally, what challenges can you see in using arts-based research in your work, and how might you overcome them?

Here's wishing you well in your own efforts to inspire creative and engaged students, scholars, and practitioners of environmental communication!

Notes

1 A more detailed treatment of arts-based research in environmental communication appears in a new book-length work (Takach 2016).

2 The tar sands have inspired high-profile, public protests in the US and Europe as well as Canada, and launched what Avery (2013) calls "the largest expression of civil disobedience since the Civil Rights movement of the 1960s."

References

Avery, S. (2013) *The pipeline and the paradigm: Keystone XL, tar sands, and the battle to defuse the carbon bomb*, Ruka, Washington.

Barone, T. and Eisner, E. W. (2012) *Arts based research*, Sage Publications, Thousand Oaks.

Boykoff, M. (2007) "From convergence to contention: United States mass media representations of anthropogenic climate science," *Transactions of the Institute of British Geographers*, 32(4): 477–89.

Brandt, D. (2006) "Introduction: Playing with wild fire: Art as activism," in Brandt, D., ed., *Wild fire: Art as activism*, Sumach, Toronto, pp. 13–22.

Canadian Association of Petroleum Producers (2010) "Canada's oil sands: Come see for yourself" [video]. Available at: www.youtube.com/watch?v=Xfl0wUXKPvo (accessed February 25, 2016).

Canadian Energy Research Institute (2014) "Canadian economic impacts of new and existing oil sands development in Alberta (2014–2038)." Available at: www.ceri.ca/images/stories/CDN_Economic_Impacts_of_New_and_Existing_Oil_Sands_Development_in_Alberta_-_November_2014_-_Final.pdf (accessed July 30, 2015).

Cardwell, F. S. and Elliott, S. J. (2013) "Making the links: Do we connect climate change with health? A qualitative case study from Canada," *BMC Public Health*, 13(208): 1–12. Available at: www.biomedcentral.com/1471-2458/13/208 (accessed July 30, 2015).

Casey, E. (2004) "Mapping the earth in works of art," in Foltz, B. V. and Frodeman, R., eds., *Rethinking nature: Essays in environmental philosophy*, Indiana University Press, Bloomington, pp. 260–69.

Corbett, J. (2006) *Communicating nature: How we create and understand environmental messages*, Island Press, Washington, DC.

Dewhurst, M. (2014) *Social justice art: A framework for activist art pedagogy*, Harvard Education Press, Cambridge.

Eisner, E. W. (1997) "The promise and perils of alternative forms of data representation," *Educational Researcher*, 26(6): 4–10.

Eisner, E. W. (2006) "Does arts-based research have a future?," *Studies in Art Education*, 48(1): 9–18.

Finley, S. (2011a) "Ecoaesthetics: Green arts at the intersection of education and social transformation," *Cultural Studies ↔ Critical Methodologies*, 11(3): 306–13.

Finley, S. (2011b) "Critical arts-based inquiry: The pedagogy and performance of a radical ethical aesthetic," in Denzin, N. K. and Lincoln, Y., eds., *The Sage Handbook of Qualitative Research* (4th edn.), Sage Publications, Thousand Oaks, pp. 435–50.

Finley, S. (2012) "Ecoaesthetics: Critical arts-based research and environmental advocacy," in Denzin, N. K. and Giardina, M. D., eds., *Qualitative inquiry and the politics of advocacy*, Left Coast, Walnut Creek, pp. 205–20.

Foust, C. R. and Murphy, W. O. (2009) "Revealing and reframing apocalyptic tragedy in global warming discourse," *Environmental Communication*, 3(2): 151–67.

Guignard, J. and Murphy, T. P., eds. (2009) *Literature, writing, and the natural world*, Cambridge Scholars, Newcastle upon Tyne.

Haseman, B. (2006) "A manifesto for performative research," *Media International Australia*, 118: 98–106.

Hodgins, P. and Thompson, P. (2011) "Taking the romance out of extraction: Contemporary Canadian artists and the subversion of the romantic/extractive gaze," *Environmental Communication*, 5(4): 393–410.

Inwood, H. (2010) "Shades of green: Growing environmentalism through art education," *Art Education*, 63(6): 33–8.

Knowles, J. G. and Cole, A. L. (2008) "Arts-informed research," in Knowles, J. G. and Cole, A. L., eds., *Handbook of the arts in qualitative research*, Sage Publications, Thousand Oak, pp. 55–70.

Lafrenière, D. and Cox, S. M. (2013) "'If you can call it a poem': Toward a framework for the assessment of arts-based works," *Qualitative Research*, 13(3): 318–36.

Leahy, S. (2006) "Oil sands: Burning energy to produce it," *Resilience*, July 27. Available at: www.resilience.org/print/2006-07-27/oil-sands-burning-energy-produce-it (accessed July 30, 2015).

Leavy, P., ed. (2015) *Method meets art: Arts-based research practice* (2nd edn.), Guilford, New York.

McKibben, B. (2005) "Imagine that: What the warming world needs now is art sweet art," Grist.org, April 22. Available at: www.grist.org/article/mckibben-imagine (accessed July 30, 2015).

McNiff, S. (2008) "Art-based research," in Knowles, J. G. and Cole, A. L., eds., *Handbook of the arts in qualitative research*, Sage Publications, Thousand Oaks, pp. 29–40.

Moser, S. C. and Dilling, L., eds. (2007) *Creating a climate for change: Communicating climate change and facilitating social change*, Cambridge University Press, Cambridge.
Pedelty, M. (2012) *Ecomusicology: Rock, folk and the environment*, Temple University Press, Philadelphia.
Quinn, T., Ploof, J. and Hochtritt, L. (2012) "It's a movement, so start moving: Art education for social justice," in Quinn, T., Ploof, J. and Hochtritt, L., eds., *Art and social justice education: Culture as commons*, Routledge, New York, pp. xviii–xxiii.
Radford, T. [dir.] (2008) *Tar sands: The selling of Alberta* [video], Canadian Broadcasting Corporation, Toronto. Broadcast on CBC Television on March 13.
Sullivan, G. (2010) *Art practice as research: Inquiry in visual arts* (2nd edn.), Sage Publications, Thousand Oaks.
Sullivan, G. (2012) "Art matters," in Quinn, T., Ploof, J. and Hochtritt, L., eds., *Art and social justice education: Culture as commons*, Routledge, New York, pp. 145–9.
Takach, G. (director) (2013) "Voices from the visual volley: Filmmakers, the tar sands and public health" [video], Geo con Brio Productions, Edmonton.
Takach, G. (2014) "Visualizing Alberta: A battle of branding, nature and bituminous sands," in Boschman, R. and Trono, M., eds., *Found in Alberta: Environmental themes for the Anthropocene*, Sir Wilfrid Laurier University Press, Waterloo, pp. 85–103.
Takach, G. (2016) *Scripting environmental communication: Oil, democracy and the sands of time and space*, Palgrave Macmillan, London.
Woo, Y. Y. J. (2008) "Engaging new audiences: Translating research into popular media," *Educational Researcher*, 37(6): 321–9.

9 Developing visual literacy skills in environmental communication

Antonio López

Visual literacy skills are essential in a society that increasingly communicates with images, especially in the realm of environmental politics and the public sphere. This is best exemplified by how political communicators use the term "optics" to designate media spin. For example, the image of President Obama visiting a melting glacier in Alaska gives optics for the challenges of climate disruption and the intention of the White House to promote policy about it. Political communicators rely on optics because "sight is by far the most trusted of our senses, and the one that has the greatest impact on our conception of reality" (Apkon 2013, 75). In this chapter I introduce a pedagogical approach to engaging visual communication that combines analytical methods developed for art history, media literacy, medium theory, and environmental communication. This multidisciplinary framework for analyzing visual communication prepares students to think critically about visual media and enables them to harness it for environmental communication.

Whether it's mass media, individual expression, social media, or alternative media, environmental communication images can be used to both hinder and promote social change, news, information, and political discourses. Choosing a form of visual communication—photography, painting, film, online multimedia, comics, writing, information graphics, etc.—requires considering the affordances and constraints of each type of visual communication within a broader media ecosystem, which I define as "the ecologically embedded sum of all our technologically mediated interactions on planet Earth" (López 2014, 27). As a whole, the media ecosystem includes "physical, sociocultural, and cognitive ecosystems, such as the lifecycle of gadgets and energy used to power the system (physical ecosystems), civic and symbolic realms (socio-cultural ecosystems), and the phenomenological experience of time, space, and place (cognitive ecosystems)" (ibid.). While the conditions for image makers in the past were constrained by a one-to-many media ecosystem that had only a few media organizations producing the majority of images in our public sphere, there is now unprecedented opportunity for a diversity of actors to increase the "symbolic action" (Cox 2009) that images can perform to promote environmental communication.

I propose that visual literacy, which is learning how to read and to communicate with visual media, can be used for increasing levels of environmental communication literacy. My approach draws from a larger literacy framework I have developed called "ecomedia literacy," which is to promote mindfulness for how everyday media practice impacts our ability to live sustainably within Earth's ecological parameters for the present and future (López 2014). This framework is designed to incorporate four perspectives that can be used for analyzing and engaging media:

- worldview (the impact of media on mindfulness, especially our perception of space, place, and time);
- environment (the material and physiological characteristics of media, in particular communications technology);
- political economy (political and economic conditions shaping communications); and
- culture (the symbolic realm shaped by media).

Within this model, the field of environmental communication primarily focuses on political economy, worldview, and culture, although not exclusively. My approach is to put all of these perspectives into conversation; however, for the purposes of this chapter, I will primarily focus on the cultural and political aspects of visual media, with the added dimension of sensory features (which relates to worldview).[1]

Overall, the approach to visual literacy presented here can be summarized by Aldus Huxley's formula for developing visual skills. As I will explain in more detail in the Visual Theories section, Huxley proposed that "seeing" combines sensing, selecting, and perceiving (Lester 2014).[2] From the standpoint of visual literacy, this means images should be understood in terms of two theoretical categories: sensory experience and meaning making. Sensory experience relates to how we are biologically predisposed to organize visual information—color, shape, size, etc.—into coherence. Meaning making is how we apply cultural experience and knowledge to our field of vision. In the following section, Environmental Iconology, I offer a detailed discussion of how iconology, an analytical technique developed in art history, can be combined with environmental communication for a system of analysis. Here I follow a three-stage process that incorporates an assortment of analytical approaches, including semiotics, environmental ideology, discourses, media literacy, and medium studies. The first stage involves examining images according to form (how they are made) and developing "active viewing." The second stage examines the cultural aspects of image construction by discussing how semiotic elements and visual grammar are used to construct meaning in images. The last stage focuses on how images are cultural maps that reflect environmental ideologies, and how to read them. The final section, Putting it All Together, offers detailed instructions for putting these theories into practice in a classroom situation. I include a sample assignment to perform an analysis of a visual environmental communication text that combines sketching and writing. The assignment, which can be a paper, presentation, or multimedia project, is designed to explore all the concepts introduced in this chapter, and provide a framework for doing a methodical and detailed image analysis. Throughout this chapter I suggest practical classroom activities that can be used to teach various concepts and theories.

As I argue elsewhere (López 2014), all media are a kind of environmental education, and increasingly this education comes in the form of images. Thus, we must cultivate the ability to mindfully engage images to critically evaluate, but also harness, visual media for environmental communication. The method of visual literacy that follows promotes fundamental skills, as the issues of our times—climate disruption, fracking, clean drinking water, sustainable food systems, etc.—are more and more defined by image-based media in the public sphere.

Visual theories: a short overview

When dealing with visual communication, it is important to understand some primary concepts about how we understand images. Broadly speaking, there are two categories of visual theories: sensory, and perceptual (Lester 2014). Sensory theories are related to how images are processed by our nervous system; perceptual theories are about how we generate meaning from images.

First and foremost, all images are created by reflected light, which has a physical property that stimulates our nervous system; our eyes "touch" light. This aligns with Huxley's concept of sensing, for it relates to how images actually stimulate our senses. One activity that can help make this point is to ask students to stare at a screen in a dark room and then close their eyes. The afterimage they will see is evidence of the light photons stimulating the cones and rods (photoreceptors) in their eyes. You can also project an inverted color image (or negative) on a screen and have students stare at it for 15 seconds, then instruct them to look away. After about 15 seconds they should see a positive version of the image floating in space (do a web search for "negative afterimage" for examples to use in class). The image is the result of the photoreceptors being overstimulated and remaining agitated to produce an afterimage.

In addition to simply responding to light, we also organize visual information according to Gestalt principles (Barnes 2012). Gestalt is a school of psychology from the late nineteenth and early twentieth centuries that examined how the mind self-organizes visual information into coherent groupings or a whole that is greater than the sum of its parts. We do this because we are uncomfortable with irregularities in our visual field, so we constantly seek out patterns, groupings, and figure/background relationships. As applied to visual awareness, this means our brain tries to make sense of visual forms (shapes, lines, dots), movement, color, and depth. In our field of vision we look at objects and organize them according to difference, similarity, proximity, continuity, figure/ground, direction, or closure (filling-in missing visual information when there is not enough to complete a picture).[3] We then compare what we see with experience to make sense of it and categorize it according to our previous knowledge. This is the "selecting" part of the Huxley equation.

The "perceiving" aspect of Huxley's approach deals with how we assign meaning to what we are sensing. This is primarily associated with semiotics. Media studies and environmental communication typically deal with this aspect of visual communication. In general, the most common semiotic methods are those developed by Ferdinand de Saussure, Charles Sanders Peirce, and Roland Barthes (Bignell 2002). These methods have in common the belief that language and images are symbolic representations and are arbitrarily assigned to objects as a result of culture. For example, chien, cane, perro, and dog are words from different languages used for a particular kind of four-legged animal that we recognize from experience. There is nothing inherent about the animal that determines the word itself; rather, the word is a human symbolic construction.

The approach I take, and one I feel is a necessary component of ecomedia literacy, is to combine both the physical (brain) and mental aspects (mind) of visual experience into our analytical method. This will help students become more aware of the embodied nature of perception as an aspect of human evolution in the physical world. To make this point, sensory techniques can be experimented with when working outside the classroom. By getting students to actively apply Gestalt principles to their outdoor visual horizon, they can also compare the same mental processes with how they look at visual media. We can also think of our eyes as not being autonomous objects connected to our brains, but in fact connected to our feet (consider how we use movement to see) and, therefore, connected "within" the environment (as opposed to separate or "outside").

Finally, drawing our attention to the sensory aspect of visual communication can open up the discussion of how screen technology impacts our physical health and also allow us to reflect on the physiological differences of a medium, such as film, TV, computers, smartphones, or print, and how they impact our brain. From the worldview perspective of ecomedia literacy, I'm also interested in how various parts of our brain process media differently, which is associated with the scholarly tradition of media ecology. The best way to open up a discussion about this topic is to simply ask students what is the difference between a photo of something (like a sunset) versus a written description. The short answer is that information in images is processed all at once, whereas writing is temporal (it takes time to read) and depends on categories for descriptions (the sun is high or low, bright or dim, etc.). Media ecology suggests the kind of media we use stimulate and develop our brain hemispheres in different ways (Abram 1996; McLuhan and Powers 1989; McLuhan 2002; Shlain 1998; Ong 1982; McGilchrist 2009). For example, the phonetic alphabet and iconographic writing can cause differences in ways the brain develops and processes visual information, such as whether we focus on specific objects or on patterns in our field of vision.

Environmental iconology

The purpose of refining and developing our analytical abilities is to identify communication patterns, and to become fluent in the lexicon and vocabulary of visual environmental communication in order to interpret and produce communication for change. As such, any media object (my term for media text) can be seen as an artifact of our age that reflects and communicates beliefs about human experience in the biosphere. To cultivate this level of awareness, I draw from an innovative approach to visual analysis called iconology, which was developed by preeminent art historian Erwin Panofsky (1962 [1939]). In combination with other analytical methods, Panofsky's iconology method can be used for exploring visual communications from their elementary form and sensory experience to semiotic readings of environmental ideologies. The technique was developed by Panofsky to analyze European visual art, such as works produced during the Renaissance, but can easily be translated to explore environmental communication in contemporary media.

Using a geology metaphor, Panofsky proposed that an image could be analyzed according to three "strata" to understand the historical moment it was produced, helping us gain insight into the cultural attitudes and beliefs that produced it. The surface stratum—the "primary" and "natural" subject matter—incorporates those aspects of visual communication that anyone could understand without prior cultural knowledge. This is principally on the level of form (*how* something is communicated) and the identification of objects. This relates to anything that can be understood based on practical experience. The second stratum—"secondary" or "conventional" subject matter—relates to those elements that require cultural knowledge from the time period the media object was produced. Just as one would need to be versed in sixteenth-century Christian theology to understand a religious painting from that period, the environmental implications of a car advertisement or an anti-fracking political campaign would necessitate awareness of current events and contemporary cultural references. To help tell the difference between the primary and secondary strata, Panofsky would apply the "ancient Greek test," which means delineating those aspects of the image an ancient Greek could understand versus those that would be incomprehensible (an ancient Greek might recognize that a car is a kind of "cart," but would not know how or why it is used). This helps differentiate between those elements that are common in human experience versus those that are culturally produced. To update this thought experiment, we could switch out the ancient Greek for an ancestral indigenous person, thereby getting students to take on the perspective of land-based cultures vis-à-vis contemporary visual cultures.

The third stratum—intrinsic meaning or content—relates to the deepest level of analysis, one in which the communicator unconsciously reflects the values and beliefs about society and the historical period in which it was created, revealing attitudes about religion, class, philosophy, and environment. Though Panofsky doesn't use the term, this can be considered the level where ideology is expressed; that is, where taken-for-granted beliefs are present in the media object. For the purpose of environmental communication, this is where one analyzes the environmental ideology of specific images, such as those found in advertising, news, activist media, or popular culture. From this stratum, one can think of images as "fictionalized" portraits that express the zeitgeist of our age (Howells and Negreiros 2012).

What follows is a detailed breakdown of how to use iconology as a framework for teaching visual literacy in environmental communication. I will also include examples of how to put these methods into practice. Each step is designed to scaffold visual analysis skills. The primary stratum focuses on form, the second deals with meaning by drawing on semiotics and visual communication language, and the third is where we explore ideology.

Primary stratum

At this level of image analysis, we use a simple method: what-you-see-is-what-you-get (Howells and Negreiros 2012). This means describing the visual elements for what they are without too much interpretation. This is a very important step, because it allows students to get into the practice of active viewing, which means consciously looking at images to seek out details, leading to one of the most important goals of visual literacy: getting students to develop the habit of looking for themselves. There are two aspects to this step:

1 to look at form, which is *how* the image is communicating (also called formalism); and
2 to identify what is present in the image.

One exercise to engage active viewing is to have students draw the image under analysis. Students can also create mind maps for what they are examining.[4] For practice, you can project images in the classroom to get students to describe what is in them. In terms of environmental communication, images that can work for this activity are eighteenth- and nineteenth-century landscape paintings by American Western painters or those produced by European Romantics. These paintings are foundational works of Western image culture that helped define contemporary discourses about the environment. This activity can familiarize students with these fundamental works and prepare them for more complex analysis.

When applying the what-you-see-is-what-you-get approach, follow these steps:

1 examine the media object according to form and style: pay attention to shapes, color, light, movement, weight, depth, lines, etc. Students also apply Gestalt principles here (how objects are organized to focus our attention).
2 Next, students should describe what is in the image, such as:
   - what genre it is (portrait, landscape, advertisement, comic, etc.);
   - the central or basic subject matter;
   - the location or setting of a particular scene;
   - the historical period depicted;
   - the season or time of year shown;
   - the time of day; and

- the particular instant depicted in the image, such as what happened before, during, and after the scene in the image (Howells and Negreiros 2012).

This kind of active viewing gets students involved in describing what is there, an important practice that is lacking in the way most people commonly engage images, especially online where the tendency is to quickly browse content.

Secondary stratum

This layer is more directly connected with cultural knowledge and deals with the image on the level of signs. In particular, this is where you can connect images with environmental discourses, which are cultural constructions. For contemporary media, this level of interpretation clearly goes beyond the purview of the ancestral indigenous person, who would not be versed in a contemporary cultural milieu. Identifying cultural codes in communication is important for many different reasons:

- it helps students understand that perceptions of living systems as reflected in visual communication are culturally constructed; and
- without realizing it, teachers and students can be ethnocentric and assume that whatever cultural practices are coded into media objects are common to all societies (Hall 2001).

Beliefs about food, for example, vary across the globe, so we can see how food culture is communicated differently in the media of diverse societies. For example, Italian food advertisements and how they depict social practices around food are very different from those of the United States. Getting students to discuss these differences can help promote awareness of the diversity of cultural experience. We can also use this as a point of discussion for how certain environmental beliefs, such as monoculture, are propagated through media. One could do a comparative study of how Monsanto is marketed in India, Latin America, and the United States to probe whether ethnocentric values are being promoted, or whether the specific cultural beliefs of a society are being reflected in a way that respects local customs and beliefs.

At this point one can introduce a discussion of semiotics, the study of signs. I assume most readers are familiar with this subject, so I only briefly review the core concepts as they relate to the framework of iconology. To keep things simple (for a topic that is fairly advanced and complex), the basic concepts I find useful in teaching semiotics combine different theories. First, students should discuss the concept of a sign, which on the simplest level is something that stands for something else. According to Saussure, all signs are composed of a signifier (what it is materially) and the signified (what it means). Film studies usually use the terms denote and connote. Again, using the what-you-see-is-what-you-get method, have students differentiate between those aspects of an image that are objective and factual with those that are culturally constructed. For example, the signifier of a green leaf in packaging could connote sustainability. Barthe's discussion of mythology, which explains why we associate dominant meanings with particular signs, can be introduced by asking students why they believe what they do about the meaning of specific signs. You could use food packaging as a way to lead this discussion, first getting students to identify the objective elements present in the package design, and then asking them what the intended meaning of those elements are (such as a label with pastoral scene on a factory farm product). What does the use of these images get us to think about the product, and why do we make these associations? The opening sequence from the documentary *Food, Inc.* (Kenner 2008) does a very good job of modeling this form of analysis.

Peirce's sign system is particularly useful. In essence, he theorized there are three kinds of signs: iconic, indexical, and symbolic (Bignell 2002). An iconic sign looks like the thing it is trying to represent (such as a lit cigarette in a no smoking poster). Indexical signs are evidence of something having occurred or connect to something not present in the image (i.e., where there is smoke, there must be fire). Symbolic signs have no resemblance to what is being represented and must be culturally learned (such as the toxic waste or nuclear symbol).[5] As will be discussed below, there are common signs used in environmental communication, such as smokestacks, polar bears, images of landscapes, Earth from space, etc., that are used to construct ecological discourses. You can have students discuss what kinds of signs these various symbols are, such as how car exhaust pollution is indexical of climate change, or how the photograph of Earth from space can be interpreted as iconic (it looks like Earth), indexical (it is evidence of Earth existing in space), or symbolic (of ecology or humanity).

Semiotics allows students to defamiliarize themselves with taken-for-granted assumptions for why signs are used and in what context. It also opens up discussion about intertextuality, the relationships between texts and how signs cross-pollinate in different media objects and across platforms. This is particularly relevant to ecopedagogy because we can start to see signs as being part of a larger symbolic ecosystem, or "semiosphere," and identify their cultural origins and how they are connected across time and space with other texts. For instance, you could compare Romanticist paintings with SUV ads to show how our views of nature are culturally constructed over time. Signs are also symbolic resources that environmental communicators rely upon to ensure the intended reading of a text or interpretation of actions such as staged image events (e.g., the Obama example at the beginning of this chapter). An example of using symbolic resources for environmental communication is the activist response to the popular film *Avatar* (Cameron 2009), which was broadly viewed around the world and widely recognized as an allegory of ecological destruction. Environmental activists from many different parts of the globe (Philippines, India, South Korea, Palestine, England) repurposed the symbol of the Na'vi (the indigenous characters in the film) to communicate their own struggles against environmentally destructive forces. When activists colored themselves blue and wore prosthetics to look like the Na'vi, they were using the symbolic resources generated by *Avatar* to connect their struggles to a familiar storyline about ecological threats. By tapping into global pop culture, activists strategically used signs to make it easier to communicate their own struggles to unfamiliar audiences.

Another way to think of signs as belonging to a symbolic ecosystem is to incorporate the concept of genre, which can be thought of as a kind of menu of expectations (Gray 2006). Film, popular culture, journalism, etc., all use sign conventions to locate themselves within particular genres, which have boundaries that are often contested and redefined. For example, Westerns (i.e., "Wild West" films) have certain defining elements, whose meanings may change as society changes. Early iterations of Westerns reflect colonialist beliefs about progress and the mastery of nature, whereas later versions reverse this to critique the destruction of indigenous culture. Returning to the example of *Avatar*, on the surface it is a sci-fi film, but it actually repurposes the Western genre to connect its ecological allegory to past struggles between European settlers and indigenous people. For better or for worse (some criticized the film for recycling the stereotype of indigenous people as romantic savages and the white European man as savior), *Avatar* tapped into preexisting sign systems to simplify its ecological message, which in turn was used by global activists to promote their own causes.

It is also important to recognize that written text and images work together. For this purpose, I encourage students to become familiar with Barthe's concepts of anchor and relay text (Gillespie and Toynbee 2006). Anchor text (such as a typical caption accompanying a newspaper photo) is used to get readers to identify something that is in an image, whereas relay text is used to get us to relate the image to something that is absent. An example of anchor text is a caption for an image of a polar bear on floating ice that reads, "A polar bear is stranded on a drifting piece of ice in the Arctic." A relay text caption would read, "As the climate warms, increasingly animals will struggle to survive in difficult and unfamiliar conditions." Relatedly, the typography of text can serve a semiotic purpose. Consider how font styles can associate media objects with particular time periods or genres (e.g., Westerns often use a particular font style in their titles). For an in-class activity, students can play with text and typography to produce cultural associations by making "anti-ads" (also referred to as "culture jamming" or "subvertisements"). An anti-ad is made by taking an existing print ad and then changing the text to subvert its original intended meaning. As an example, on the Internet an unknown author converted a print ad that depicts an SUV rampaging in a desert landscape with a revised caption that states, "Nature … IT'LL GROW BACK." Another anti-ad for a Cadillac SUV is revised to read, "Catastrophic climate change? Not our fault." (These are viewable along with other anti-ads at https://it.pinterest.com/proflopez/culture-jamming/). As an example of anti-ads being a form of environmental communication, during the United Nations 21st Conference of Parties climate negotiations in Paris in 2015, the Brandalism Project (www.brandalism.org.uk/) installed over 600 artworks in traditional ad spaces (such as bus stops) in an effort to subvert corporate influence on the talks. For a class activity, I use the

Bubble Project (www.thebubbleproject.com/), which is a template of blank speech balloons that can be applied to ads. Students are challenged to write different speech balloons that are then applied to existing images that demonstrate anchor and relay text.

An important component of reading images is also recognizing the significance of visual grammar. As the term suggests, grammar is how signs work together to produce new meanings in the same way isolated words are combined to create sentences. In the case of images, signs can also be visual elements such as camera angles or frames on comic book pages. Significantly, every visual medium has different grammatical rules. For example, classic Hollywood cinema has visual conventions for telling stories that are different from films produced in other cultures, such as those in Japan or India. But film itself has a recognizable visual grammar that is different from other types of media, such as magazines or comic books. Any analysis should take into consideration how medium formats use different grammar because, when evaluating or planning environmental communication, one must consider the appropriate affordances or constraints of the medium. For instance, comics use speech balloons with squiggly lines to indicate internal thoughts; a dissolve edit in film indicating a flashback; and TV news uses presenters, crawls (text scrolling on the bottom of the screen), and brightly colored alerts to indicate "this is news." You can compare and contrast how image campaigns are communicated across different platforms. Environmental communicators such as Greenpeace run media campaigns in different formats, thus they must consider the appropriate visual grammar available for their different choices. A 30-second public service ad for TV has a range of choices that differ from a print advertisement or billboard.

To summarize, having students practice semiotic reading cultivates awareness of how meaning is constructed. For the purpose of environmental communication, the semiotic approach scaffolds more complex levels of analysis, in which students explore the concept of metaphors and tropes that make up discourses. Because environmental communication is interdisciplinary, there are different conventions used to describe discourse construction; I will assume the reader can apply his or her preferred method to the secondary strata. In short, the grammar of discourses like "progress," "eco-apocalypse," "eco-utopia," "pastoral," "cornucopia," "wild/wilderness," "pollution," etc., are comprised of signs, and the ability to read signs will then help students make sense of discourse construction and make connections between discourses and environmental ideology, which is discussed in the next subsection.

Third stratum–intrinsic level

On the intrinsic level, the analysis draws on Corbett's (2006) discussion of environmental ideology, which is one's beliefs of how humans should interact with living systems. Like Panofsky's discussion of art, all media embed beliefs that are products of the historical moment that produced them (hence they are cultural "maps"). In environmental communication, students are trying to identify taken-for-granted beliefs about the relationship between humans and living systems that are represented in media. Before entering this stage, students should already be familiar with the spectrum of environmental ideologies, which ranges from anthropocentric (human-centered) to ecocentric (ecologically integrated). Usually I have students discuss at least one media object for each environmental position Corbett places on the spectrum (such as unrestrained instrumentalism, conservationism, preservationism, ecofeminism, deep ecology, and indigenous traditions), and then compare and contrast them while asking students to identify and debate which ideology is represented in the media object. The examples include TV advertisements, film clips, political ads, documentary clips, public service announcements, and news segments (follow this link for a YouTube playlist I created with examples to use in class: http://bit.ly/1PeX1QZ). Other examples can vary from print to moving images to web sites.

As a rule of thumb, try to use problematic media, such as segments from the animated film of the Dr. Seuss book, *The Lorax* (Renaud and Balda 2012). On the one hand, there are ecocentric discourses present in the movie, but then the film was cross-promoted in Mazda's CX-5 SUV advertisement, which seems to undermine the film's environmental messaging. The fast food restaurant chain Chipotle has produced a number of animated videos that promote sustainable food production, but the chain could be considered to have an ambiguous ecological position that can be problematized. The main thing is to get students to consider to what extent the environmental ideologies being expressed are taken for granted or are consciously constructed (i.e., SUV ads have taken-for-granted assumptions that the "environment" is a playground or a place isolated from humans). Advocacy media promoting environmental causes, in particular by radical environmentalists, can be seen as counter narratives to dominant ideologies and therefore would be examples of self-conscious ecocriticism intended to disrupt the symbolic order (DeLuca 1999).

Finally, an often-neglected step in any image analysis is a critical assessment of the communicator or media organization that produced the media object. Students should be encouraged to look at the mission statement of the communicator to determine how the organization is connected to a particular environmental ideology being expressed. For example, students can compare and contrast the mission statements of ExxonMobil and Greenpeace by researching their respective websites. This is a productive activity because students will discover that finding a mission statement (or corporate responsibility statement, as the case may be) is not always easy, which reflects upon the level of transparency (and hence accountability) of the communicator. Just as importantly, the difference between private, public, for-profit, non-profit, audience-supported or state-sponsored media organizations has a direct impact on the ideologies communicated in the media they produce and connects the media object to larger economic and social practices. For example, it is useful to compare the mission statements of for-profit news organizations such as Fox News Channel with those of audience-supported media such as Democracy Now! and then explore ways environmental issues are covered differently by these organizations.

Finally, a comment on classroom technique. Up to this point, the three-strata process should be modeled and explored with students so they become familiar with the main concepts. To spark discussion in the classroom, I use three examples for each stratum. The reason is not scientific; it just seems to have the right level of redundancy to get the concepts to stick. I find it productive to put two media objects side by side to compare and contrast, as the juxtapositions tend to yield interesting discussions. I also ask students to consider media objects as cultural maps; I encourage students to think of themselves as alien anthropologists to try to get them to defamiliarize themselves with taken-for-granted beliefs embedded in media objects. This helps create some distance from their own assumptions and familiarity with the media around them.

In the next section I discuss an image analysis assignment that students perform on their own, using a formula derived from media literacy practices and refined for the purpose of teaching environmental communication.

Putting it all together

Students may write a paper, give a group presentation, produce a video, or create an annotated multimedia project. For the assignment, it is effective to have them analyze a magazine cover, a movie poster, or an advertisement, because they are simple to study and combine a maxim amount of elements to be analyzed (e.g., typography, text, photography, color, Gestalt). You can certainly have them analyze other types of image, such as art or photojournalism, but I find those examples lack enough elements for an extended in-depth analysis. Media objects comprising multimedia or moving images have many more components that can make the assignment overly complex, and therefore should only be considered for more advanced courses. Before they write or present, encourage students to make different lists using the analytical tools listed below and then try to connect them visually with a mind map. They should also draw the image they will analyze, which is common practice in art history. It allows for deeper viewing, enabling students to experience sensually the form and composition of the image. Additionally, students can use free online tools, such as ThingLink (www.thinglink.com) or Explain Anything (http://explaineverything.com), to annotate their images and share them on the web.

In your assignment prompt, inform students their analysis should incorporate the following theoretical and analytical tools following the structure of iconology:

First stratum

- What-you-see-is-what-you-get: make an inventory of what is in the image (noting unique compositional elements).
- Formalism: note visual cues of color, form, depth, and movement.
- Examine the image based on Gestalt laws and sensory experience.
- Graphic design: examine typography and design elements, such as contrast, size, balance, rhythm, and unity.

Second stratum

- Identify iconic, indexical, and symbolic signs.
- Examine the use of words (anchor or relay text) and how they influence meaning.
- Ascertain any issues related to cultural codes in regard to environmental representation, human-nature relations, and racial/cultural/gender stereotyping.

Third stratum

- Identify signs/tropes/symbols connected to environmental discourses.
- Inventory environmental ideology position/s.

It should be emphasized that the image sketch and mind map are tools to gather information and formulate themes in preparation for the final analysis (I don't ask students to turn in these materials because I don't want them to feel the pressure of a grade during brainstorming activities). They should then use these materials as a basis for the next step, which is formulating an analysis based on a series of questions listed below. Please note that students might feel overwhelmed by the various theoretical lenses used, so it is important for them to understand that each analytical method should be considered a tool for doing different things. I use the metaphor of a Swiss Army knife to explain that when we need to perform a function (like open a bottle), we don't use every tool on the knife, but rather select the right one for the appropriate task. Additionally, they should be encouraged to use Google's Image Search tool, which can be used to find background information on any image (follow this link, https://images.google.com, for instructions on how to submit images to the search feature).

Using a structured technique based on media literacy analysis practices, students now follow a series of questions that allow them to enter into dialogue with the media object. The subsequent questions are used as a guide for building a structured argument. In their final project, I require them to communicate a narrative presenting their analysis with an introduction, transitions, and conclusion without simply composing it as a series of questions and answers:

- *How is the media object represented?* What is its presentational context, such as magazine type, TV channel, or website? Is it from a screen-based media gadget, and, if so, what kind and what format? Are there any other medium properties worth noting? Does this physical context influence its meaning?
- *What is the impact of the media object on sensory experience?* How is it expressed in terms of Gestalt laws, and what is its affect on your experience of time, space, and place?
- *Who is the communicator?* Is it an individual, NGO, corporation, state, or other kind of organization? What type of producer is it (private, for-profit, nonprofit, public, audience supported)? Is the communicator authentic?

How do you know? Include some background information about the communicator and reference sources.

- *Who is being targeted?* Each image tells a story, so to whom is the story told? Who is the intended audience? How do you know?
- *Why did the communicator produce this?* What are the motives of the communicator, and what is the purpose and function of the media object?
- *How was it made?* What is its format? What do you see/hear? What is the story? What symbolic resources does it use? Does it express a particular media grammar?
- *How does the media object express environmental ideologies? What does it say; and what does it* ***mean****?* Is it fact, opinion, persuasion, or something else?
- *What kinds of ecological values are being expressed?* How does it orient to your own worldview, family experience, personal ethics, cultural identity, and social status?
- *From an environmental communication framework, is the media object accurate, fair, truthful, and complete?* Is there anything left out? Is it real and true to life? Does it benefit or harm anyone or ecological element, process, or system?

The final assignment should demonstrate some fluency in the various concepts covered in this chapter. Educators are encouraged to develop their own approach and modify the assignment according to the needs of a specific course setting. Ultimately, what is described here has been tested dozens of times and invariably allows students to put into practice the theoretical and analytical methods covered.

Conclusion

Given the shift to image-based communication in social media and smartphone use, the study of environmental communication should emphasize and develop further visual literacy skills so we all can learn to perceive as Aldus Huxley had hoped, which is to see the world more deeply. As such, an ecoliteracy model of visual communication combines the study of visual languages, technological properties, and the material reality of media devices. In this chapter I have primarily focused on visual characteristics by building on the iconology framework for environmental communication. Iconology's three-layered process enables students to explore form, symbolism, and ideology by developing the following key visual literacy skills:

- learning to actively see;
- understanding visual language by recognizing the significance of cultural knowledge as a component of communication; and
- identifying how environmental ideologies are communicated.

Along with other approaches to environmental communication, visual literacy is a substantial feature of ecomedia literacy, for it encourages critically engaging the image environment that shapes human interaction with living systems, both in terms of conveying environmental ideology and promoting an awareness for how to use imagery to solve our ecological crisis. The framework presented in this chapter promotes the awareness that images function within symbolic ecosystems and also belong to particular medium ecosystems with specific grammars and technological affordances. Finally, the method offered here teaches the significance of sensory awareness, connecting imagery to the biological condition of our physical experience in the world.

Notes

1 It's worthwhile to connect visual media with the material aspects of media technologies, in particular the environmental impact of screen-based media, but that discussion is beyond the scope of this essay. See Maxwell and Miller's (2012) *Greening the Media* and the essay collection Maxwell, Vestberg and Raundalen (2015) *Media and The Ecological Crisis* for an excellent introduction to impact of media on environmental and specifically human health.
2 Rather then cite primary sources, I'm mainly citing introductory texts that offer overviews of these concepts so that the bibliography can serve as an introductory list for teachers and students.
3 For examples of each of these Gestalt principles you can do an Internet image search using the following key terms in combination with Gestalt: similarity, proximity, continuity, figure/ground, and closure. You can also visit the Pinterest board I created for this concept: https://it.pinterest.com/proflopez/visual-elements-intro-to-visual-communication/.
4 Mind maps are diagrams that can be drawn on paper or made by using free software available on the web (a simple internet search will reveal a number of easy to use tools). It starts with drawing a circle around a central concept at the center, and then drawing branches to other concepts, and linking between them with lines to visually make connections between concepts. Mind maps can also be more creative by drawing imagery to represent concepts.
5 As a side discussion, you might ask students how people in the future will know what these symbols mean.

References

Abram, D. (1996) *The spell of the sensuous: Perception and language in a more-than-human world*, Pantheon Books, New York.

Apkon, S. (2013) *The age of the image: Redefining literacy in a world of screens*, Farrar, Straus and Giroux, New York.

Barnes, S. B. (2012) *An introduction to visual communication: From cave art to second life*, Peter Lang, New York.

Bignell, J. (2002) *Media semiotics: An introduction* (2nd edn.), Manchester University Press, Manchester.

Cameron, J. (2009) *Avatar* [film], Twentieth Century Fox.

Corbett, J. B. (2006) *Communicating nature: How we create and understand environmental messages*, Island Press, Washington, DC.

Cox, R. (2009) *Environmental communication and the public sphere* (2nd edn.), Sage Publications, Thousand Oaks.

DeLuca, K. M. (1999) *Image politics: The new rhetoric of environmental activism*, Guilford Press, New York.

Gillespie, M. and Toynbee, J., eds., (2006) *Analysing media texts*, Open University Press, Maidenhead.

Gray. J. (2006) *Watching with The Simpsons: Television, parody, and intertextuality*, Routledge, New York.

Hall, S. (2001) "Encoding/Decoding," in Durham, M. G. and Kellner, D., eds., *Media and Cultural Studies: Keyworks* (rev. edn.), Blackwell, Malden, pp. 166–76.

Howells, R. and Negreiros, J. (2012) *Visual culture*, Polity Press, Cambridge.

Kenner, R. (2008) *Food, Inc.* [film], Magnolia Pictures.

Lester, P. M. (2014) *Visual communication: Images with messages*, Wadsworth/Cengage Learning, Boston.

López, A. (2014) *Greening media education: Bridging media literacy with green cultural citizenship*, Peter Lang, New York.

McGilchrist, I. (2009) *The master and his emissary: The divided brain and the making of the Western world*, Yale University Press, New Haven.

McLuhan, M. (2002) *The Gutenberg galaxy: The making of typographic man*, University of Toronto Press, Toronto.

McLuhan, M. and Powers, B. R. (1989) *The global village: Transformations in world life and media in the 21st century*, Oxford University Press, Oxford.

Maxwell, R. and Miller, T. (2012) *Greening the media*, Oxford University Press, New York.

Maxwell, R., Vestberg, N. L. and Raundalen, R., eds., (2015) *Media and the ecological crisis*, Routledge, New York.

Ong, W. J. (1982) *Orality and literacy: The technologizing of the word*, Methuen, London.

Panofsky, E. (1962) *Studies in iconology: Humanistic themes in the art of the Renaissance*, Harper & Row, New York.

Renaud, C. and Balda, K. (2012) *The Lorax* [film], Universal Pictures.

Shlain, L. (1998) *The alphabet versus the goddess: The conflict between word and image*, Viking, New York.

10 Designing a distance education course in environmental journalism

Gabi Mocatta

In the last half century, awareness of human impact on our planet has grown to such an extent that environmental debates have a permanent, central place in public discourse. With the increasing demand for expertise in communication on the environment, courses of study that rapidly deliver a thorough overview of environmental communication (EC), along with an in-depth theoretical and practical grounding in its specific sub-disciplines, including environmental journalism (EJ), are vital. If "the environment" is a social construct, as much research confirms, then educating future environmental communicators to produce and transform that construct in an informed and ethical manner plays an important role in steering the future of our planet.

Environmental communication studies, like most fields of teaching and learning today, are subject to the trend toward flexible, self-directed, and life-long learning. This means EC education is no longer confined to lecture halls and tutorial rooms of universities but has also expanded to online contexts. Non-university institutions now offer EC courses, many taught wholly or partly online. Although there is an established literature on online pedagogy, little work has been published specifically on teaching any aspect of EC online or by distance education. This chapter aims to help fill that gap.

The discussion here, therefore, analyzes the design of a distance education course in environmental journalism to be offered by a new online journalism school as part of a professional certificate in journalism. The course is written and will be tutored online by this author, who has previously taught online, partly online, and face-to-face university courses in journalism, media, and communications. I argue here it is not only the instruction itself but also the pedagogical model that informs the teaching material that is a crucial consideration in EC course design. This may be particularly so for distance or online education offerings, which, if designed well, can have an advantage over lecture-style, face-to-face courses by encouraging learners' own construction of knowledge, thereby enhancing their capacity for real-world application of new learning. In this context, this study seeks to show how competent course design might allow EJ to be effectively taught and learned online, addressing many of the concepts and theories that inform the field of EC as a whole.

This chapter first establishes how EJ fits within the EC discipline, then turns to some of the theoretical background of online pedagogy, placing it within the context of distance education. Then, using the present EJ course as a case study, I address instructional design and provide sample content to illustrate different facets of EJ in a distance and online setting. Finally, this chapter also assesses some aspects of distance study that may be *less* effective for teaching EJ.

Journalism on the environment as environmental communication

Once an emerging discipline, EC now has behind it some thirty years of scholarship. Whereas three decades ago we might have had to explain just what was meant by "environmental communication," the field is established enough today that we can identify within it a number of discrete spheres. "Green" marketing, environmental protest, public relations, communicating risk, cultural environmental meaning systems, and EJ, for example, constitute some of the key facets of EC. All these aspects of environmental discourse—public and private, organizational, governmental, mediated, and one to one—are strongly interconnected. As one of these facets, EJ cannot be considered in isolation from EC as a whole.

EC scholarship tells us "the environment" and human responses to it are constructed by contemporary and historic events and acts, images, protest, media reporting, and, importantly, the *framing* of these representations (Hansen 1991, Anderson 1997, Lester 2010, Hansen and Cox 2015). This means EC is fundamentally informed by the social systems in which it is embedded, and subject to power structures within those systems. Social, economic, political, and ideological forces all constrain or allow the ways in which "the environment" is communicated, making EC a symbolic, constructive medium (Cox 2012) and a "social operation so powerful that it changes mental states" (Cantrill and Oravec 1996, 11).

Journalism on the environment is at the nexus of these sites of social construction, symbolism, and struggle for power. In an increasingly conglomerated global media landscape, the corporate news media both reinforce the prevailing hegemony and can be an arena for *contestation* of power (Cottle 2006). Contesting actors in mediatized environmental debates compete for the symbolic power (Bourdieu 1991, Thompson 1995) of constructing "the environment," defining environmental problems, and influencing responses to them. This means journalists who report on the environment are the focal point for a variety of EC output. Public relations staff from government, environmental advocacy groups, and business all aim to persuade and influence journalists with their information subsidy on the environment. These environmental communicators may indeed be environmental journalists' main sources, so that environmental journalists must have an in-depth understanding of their roles and motivations.

Environmental journalists, therefore, work at the core of environmental debates and hold arguably the key role in the spectrum of EC. Environmental journalists interpret, filter, and curate EC output both by environmental movements and by organizations that have the most powerful environmental impacts. With its potential to influence policy-making that affects the state of our planet, EJ is inherently political. This means journalists who write on the environment are at the heart of the advocacy debate in EC. Environmental journalists must ask themselves whether their practice should align itself with pro-environmental politics, and its attempts to attribute value to the natural environment, often contrary to the interests of power, wealth, and technology, of which the corporate media today form a key part. Environmental journalists must likewise question whether they have the *ethical* duty that Cox ascribes to EC as a whole, to "enhance the ability of society to respond appropriately to environmental signals relevant to the well-being of both human communities and natural biological systems" (Cox 2007, 5). All these factors demand of an environmental journalist skillful navigation of a complex suite of socio-political, technical, and natural phenomena, making good EJ a formidable task.

But for a journalist to cover the environment exclusively is rare indeed. With the current reduction in hard news coverage by most mainstream media outlets, few media organizations now have a dedicated environmental beat. Research shows, in the US at least, that stories on the environment may be covered by science writers, agriculture specialists, health reporters, or simply by generalist journalists (Sachsman et al. 2008). This is to the detriment of a field that demands expert knowledge and long-term experience of complex, specific issues and their political and social contexts. So at a time when EJ has become more important than ever, "the ranks of reporters best equipped to cover … major environmental and climate change stories at most news outlets, particularly in local markets, are being decimated" (Daley 2010).

For those with an interest in studying EJ and reporting on the environment, this tendency indicates the need to cultivate *flexibility* above all. Though numbers of environmental journalists may be reducing, studies show an increasing demand for better information, with 79 percent of Americans polled as wanting improved environmental news coverage (PIEC 2015). At the same time, new models and partnerships beyond mainstream media, such as crowd-funded journalism, non-profit news outlets, topic-focused EC institutions, and university-based research reporting collaborations, are opening up for environmental news. Contemporary environment reporters may, therefore, find themselves communicating not just through traditional media outlets but by means of any of these emerging models. They are also likely to spend their careers working alternately in EJ positions and in any of the many other roles that constitute EC as a whole. On a professional culture level, therefore, EJ and EC—and the people who do them—are closely linked and sometimes even indistinguishable. It is precisely this mobility between EC sub-disciplines that means anyone with an interest in any aspect of communicating on the environment would be well served by studying EJ.

Just as the shift in the global communications paradigm brought about by the Network Society (Castells 1996) has resulted in the kind of profound changes to news media noted above, study and learning also have been influenced by this powerful configuration principle. In practice, this means an increased demand for online, flexible, learner-directed courses that offer opportunities for lifelong study. It is to this new learning paradigm I now turn.

Distance and online pedagogy in theory and practice

As a field of pedagogy, distance education is well established. Research has identified five "generations" in the development to date of distance learning, originating with correspondence study, and leading to today's interactive multimedia, eLearning courses taught entirely online (Taylor 2002, M. Moore and Kearsley 2011) at times with computer automated responses to students, for example, the automatic essay scoring (AES) systems sometimes used in massive open online courses (or MOOCs) (Reilly et al. 2014). Terminology in the distance education field to date has been applied rather loosely, to the extent that studies have been made of the differing meanings—and corresponding student expectations—of terms like eLearning, distance learning, blended learning, and online learning (J. Moore et al. 2011).

Distance education is defined in this chapter as any course of study where the student is geographically distant from the location of the instructor. Distance learning is now almost exclusively led by online instruction (eLearning) and supported by online material, though instruction may also include hard copy texts (hybrid learning), face to face (or blended) and virtual face-to-face learning components. Distance learning in practice has historically been strongly influenced by the dominant prevailing model of pedagogy, the long-established *objectivist* (or transmission) model, where learning is assumed to be a one-way transmission of knowledge from teacher to student and students' understanding is measured by examining their ability to produce a predetermined correct response at assessment (Lakoff 1987, Jonassen 1992). In this model, interactivity and collaboration are limited. The objectivist model also assumes a transmission of values and ideologies from instructor to student. This makes it particularly easy for a student to reproduce the instructor's position on the given subject, rather than developing her or his own views. In a field as symbolic and value-driven as communication on the environment, this has important implications. While instructors who come from an environmental advocacy standpoint might prefer students to adopt their viewpoint, courses that allow students to form their *own* opinions may ultimately create environmental communicators who advocate for the planet with more conviction. Creating learning content and experiences that facilitate students' formation of their own opinions and positions is a priority of the constructivist pedagogical model.

In the *constructivist* concept of education, most strongly associated with Piaget (1970), Kuhn (1996), and Vygotsky (1978), knowledge and meaning are constructed in the mind of the learner, and also in communities of practice, through social interaction. In constructivist pedagogy, there is no one truth or reality, meaning that education must encourage multiple perspectives. Learners make their own interpretations of information. The priority of the constructivist approach is therefore to cultivate learners' thinking and knowledge construction skills, to encourage interaction with practitioners in the field of study and also to encourage collaboration between students themselves as a "community of learners" (Vrasidas and Glass 2004). The constructivist model is therefore an exercise in meaning making, in which each individual uses prior knowledge to create his or her own unique response to any learning material provided. As a pedagogical model, constructivist learning would seem to be well suited to teaching EJ online, allowing a globally disparate student cohort to construct—and, with fellow students/instructors, co-construct—the type of EJ most relevant to their own context.

In teaching EC/EJ—or almost any course of study—higher education institutions, by contrast, tend to run on the traditional, objectivist model of face-to-face "expert" lectures. This is perhaps chiefly because of a high student-to-faculty ratio and financial constraints on lowering this ratio. But online technologies may reduce the cost of teaching using a constructivist model. There is now some evidence to suggest that, rather than being a passive, solitary mode of study, the online/distance study medium may be inherently compatible with active, collaborative, constructivist learning (Chen 2007, Schell and Janicki 2012), promoting successful learning outcomes. A recent meta-analysis of fifty-one studies of online learning found that courses which make at least some use of online learning produce more favorable student learning outcomes than classes using solely face-to-face instruction (Means et al. 2009). Of course, just making a course of study available online by no means guarantees learning success or a constructivist learning experience. Student outcomes for *wholly* online courses with no built-in interaction or collaboration were similar to those for traditional, lecture-based *wholly* face-to-face courses (ibid.). It is therefore not the medium *itself* that conveys the advantage in online learning, but other characteristics, such as its capacity for co-creation of learning, the expansion of learning time that online courses have been found to encourage (ibid., 18) and, as I argue here, the style of pedagogy to which the online medium lends itself.

It seems that some structural features of online learning *do* inherently support the kind of constructivist pedagogy that is suitable for teaching EJ. Self-managed study allows learners control over the learning process which research demonstrates is an important factor in constructivist learning success (Zhang 2005). By requiring self-managed study, courses such as this EJ offering with an online component can also develop students' capacity for active learning (Palocsay and Stevens 2008), promoting problem-solving, reflection, critical analysis and higher-order thinking. And, with a built-in requirement for collaborative knowledge construction, online study can support the "community of learners" notion that is critical to constructivist pedagogy. It has been shown, in addition, that, in comparison to face-to-face courses, online courses have a higher level of student-initiated communication with their instructors (Arbaugh and Rau 2007). Instructors can extend this, given time and motivation, to ensure it becomes the kind of interaction with experts in the field that is a mainstay of constructive pedagogy. Distance/online learning has many positives to offer EJ students without many of the time and geographical constraints associated with in-classroom study.

On a practical level, constructivist distance learning of EJ or many other subjects can include online web rooms or video conferences—"virtual" face-to-face components—where students/instructor interact in real time. Asynchronous communication between instructors and learners can take the form of message boards, discussion forums, or dedicated social networking pages. While these also support the community of learners concept, the fact that communication here is carried out asynchronously is important as asynchronous discourse may encourage more self-reflection and thus deeper learning than synchronous discourse (Hiltz and Goldman 2005; Jaffe et al. 2006).

Of course, teaching and learning online has drawbacks. For disciplines that involve practice, such as EJ, the necessary back-and-forth interaction and the individual, nuanced feedback of the face-to-face classroom may be difficult to replicate online. Also, the larger the group size in synchronous online work, the more difficult it may be to manage meaningful collaboration. A large instructor-facilitated EJ study group may tend away from online constructivism and toward transmission-style teaching, by default.

Although they differ so fundamentally, the objectivist and constructivist models can be thought of as a continuum (Vrasidas 2000). This means any teaching—whether of EJ or other subjects, whether distance or face-to-face—often blends the two approaches, even within one course, dependent on the context, student demographic, learning objectives, and technologies available. Although the distance education course I use as the case study here emphasizes constructive learning, the course draws on aspects of both pedagogical models outlined above to facilitate students' learning of EJ.

Designing a distance course in environmental journalism: a case study

The school

Open School of Journalism (OSJ) is an ISO 29990 certified independent, international school of journalism. The school is a typical internet "startup," run out of offices in New York and Berlin, with courses due to commence in 2016. The school will offer sixty courses via distance education, much of each delivered online. Courses are written in English by media practitioners and academics from thirteen countries. Courses cover both core journalism skills and specific genres or specialist areas of journalism, including EJ. Students will select a range of modules for between twelve and twenty-four months of study. On completion, students will receive an ISCED (International Standard Classification of Education) Level 4 pre-tertiary professional education Certificate in Journalism, which is internationally recognized by UNESCO.

Who studies by distance/online … and why it matters for EC, EJ and the planet

The projected OSJ student demographic is typical for distance students today. Students tend to be older than those studying in campus-based programs: 45 percent are under 30 years, with most of the remaining 55 percent between 30 and 50. Well over half are in full-time employment, and 83 percent of this demographic already have a graduate degree. Interestingly, in the context of this course, the school's target demographic seems to correspond closely with the demographic that research suggests is the most environmentally concerned: those with higher levels of education and higher incomes (Franzen and Vogl 2013). The age groups of generation X (30–44) and Y (less than 30)—the bulk of OSJ's intended student cohort—also show high levels of environmental concern on some indicators (Newman and Fernandes 2015). This trend is important as it suggests where students embarking on a course in EJ may *already* stand in their approach to environmental issues and has implications for their likely response to ideas and values revealed in this course. This was strongly borne in mind during the creation of course content.

Perhaps the most significant aspect of this course is that it is not a university study offering. This means the entry barriers for students are lower, as is the cost. The content and language may also be more accessible than they might be in a university module on any aspect of EC. All this is arguably good for communication on the environment because it means that the study of EC and EJ have truly "gone mainstream." The more people that study any aspect of EC and learn to do it *well*—from an ethical, advocacy standpoint and as agents of positive environmental change—the better it is for the planet.

A hybrid print/online learning environment

OSJ refers to its offerings as "distance education" (rather than simply "online" courses) because each course has a hard copy course text, written by the course author, situating the course in a print/online hybrid environment. A printed, posted course text might seem to hark back to the days of correspondence courses by mail—and is clearly disadvantageous from an environmental sustainability viewpoint. However, research indicates that comprehension from print reading may be higher than that from reading digitally (Mangen et al. 2013) and that for essential readings, students overwhelmingly prefer print, which is "more effective for deep engagement and learning" (Mizrachi 2014, 736).

For online EJ learning at OSJ, students will join weekly instructor-led audio/video supported virtual web rooms using the open source online conferencing system BigBlueButton. This system enables document sharing, chat, shared desktop viewing, use of PowerPoint slides, an annotatable whiteboard, and recording for later playback. Here, students will be able to engage in synchronous, constructivist EJ learning with their colleagues and instructor in virtual face-to-face format in real time. Of course, synchronous online learning for a truly global cohort is complicated by time zones, and this will need to be a consideration of this or any similarly international online course. Less time-dependent, clearly, is asynchronous online interaction and OSJ students will also be able to collaborate with colleagues in asynchronous, unmoderated learning forums. As every instructor who has taught online knows, however, participating online is not automatic amongst students signed up for distance study. Ideally, a course needs some built-in impetus to get online. In this EJ course, students are asked to submit end-of-chapter and end-of-course assessments online and can receive individualized tutor feedback fortnightly in the course's online portal.

Composing a hybrid course that would offer print-based opportunities for active learning of EJ, as well as a strong impetus for online constructivist collaboration, was part of the design-challenge of this course. Online instructors might agree that providing students with rich content relevant to *them* is crucial to stimulating online engagement. I therefore now turn to a discussion of course content to show how this has been designed to respond to the challenges—and leverage the advantages—of learning EJ online.

Designing EJ content for the distance medium

One feature of a non-university distance education course is that, with no entrance exams or standard entry qualification, the student cohort may be highly heterogeneous. A specific design challenge, therefore, comes in knowing at what level to pitch content. Students of this course were assumed to have some tertiary education and basic environmental knowledge, though not necessarily much grounding in journalism (the course can be one of students' first or early courses at OSJ). A specific advantage of non-university distance learning, on the other hand, is the flexibility afforded by not having to fit material into a thirteen-week semester or into any pre-determined, time-dictated format. This means a course can quickly deliver highly tailored content, covering only those theories and concepts most relevant to a very specific subject matter, enabling all students to participate on the same level, no matter what their backgrounds.

This course, therefore, provides a select primer in media theory relevant to EJ, and covers key concepts and themes in EC as well as practical journalism skills. Students are asked to consider initially what EJ is and what an environmental journalist does, and then to understand EJ's role in the EC field. Students investigate concepts such as the environment as a construction, the mediatized contest for environmental issue definition, the objectivity/advocacy debate, communicating risk and crisis, and visual rhetoric in EC/EJ. Students are also required to produce practical journalistic output with the potential for publication, though having material published is not a requirement of this course.

Another hallmark of distance learning is that students take many study breaks. Distance courses therefore need a particularly clear structure to reorientate students within the learning material. To this end, this course has overall learning outcomes, outcomes for each course section, a review and prompted reflection at the end of each chapter, end-of-section assessments, and an exam which asks students to build on each section's learning outcomes, with subsequent instructor feedback—another opportunity for students to hear from a practitioner of EC and EJ.

To assist in providing orientation as students "dip" into and out of learning, the course also has a dominant and recurring theme, or *leitmotiv*: the advocacy debate in EC/EJ. The course attempts to reveal the conflicted position in which environmental journalists find themselves as communicators who construct "the environment." Students are prompted to ask whether environmental journalists can and should be advocates for the biosphere or whether they are inextricably bound by the imperatives and values of the corporate news media. In alignment with the constructivist pedagogical model this course prioritizes, students are challenged to establish their own position on this. Although no "correct" answer is provided, the course *does* aim to highlight the potential for environmental journalists to be transformative agents for restorative environmental change. Tasks and questions in the assessment sections both implicitly and explicitly help students think through the advocacy/objectivity debate—prompting them to reflect deeply about what kind of environmental communicators they might themselves become. The structural feature of a strong theme or *leitmotiv* designed to provide a clear study framework for distance students therefore also serves to enrich content and enhance the constructive learning experience.

With place and local knowledge constituting key themes in EC, the course content was also designed to leverage the importance of place. The course text is explicated with numerous local examples, which, as course author, I chose particularly to provide an international focus, given that the student cohort will be globally dispersed. For example, by including illustrative examples from the Global South, the course aims to be accessible and relevant to students in a variety of cultural contexts that have varying concepts of place, "the environment," and the role of the journalist. But being in some ways "placeless" can also be a distinguishing advantage of learning online. Using the constructivist model, the course asks students to bring their *own* understandings of place and their *own* country's environmental context to their studies (rather than imposing this, objectivist style). To apply and synthesize new knowledge, for example, students are asked to select and research real-world case studies of environmental issues that affect them. Sample professional output they produce must demonstrate their own unique local environmental insights. The content is also illustrated with several examples drawn from Australia's ecocultural context, revealing the course author's location. As author, I felt that by identifying and analyzing local environmental issues that directly affect or concern *me*, I could invite students to think about issues local to *them*, enabling them to apply learning from the course to situations they know best.

Applied constructivism: a sample EJ practice task

One example of constructivist content that requires students both to consider their stance on the advocacy/objectivity debate and that also leverages the global/local dichotomy of place is an image-making exercise. Building on their theoretical study of symbolic environmental imagery and practical discussions on documentary-style image-making in this course, students are asked to explain why environmental images are often abstract, decontextualized, and symbolic with reference to Figures 10.1–3.

This part of the exercise does use the objectivist model, testing students' ability to recall new material covered in the course. Importantly, though, this is used as the foundation for an image-making task on the constructivist learning model. Students are asked, using any camera, even a mobile phone—to create:

- a symbolic stock image-type environmental photograph;
- an image that could be polysemic;
- an image with an exclusionary frame—where the frame may have changed the meaning of the environment image; and
- four documentary-type photos related to the environment and an extended (50 word) caption for each.

With the convergence of digital image-making and online platforms, creating sample imagery is a study task particularly suited to both synchronous and asynchronous online peer feedback and collaboration. In subsequent online sessions, students will be expected to present and critique imagery with their community of learners, with the aim of convincing other students of their own environmental

Figure 10.1 Sample EJ practice task 1.

Source: worradmu, freedigitalphotos.net.

Figure 10.2 Sample EJ practice task 2.
Source: © Larissa Beumer/Greenpeace.

Figure 10.3 Sample EJ practice task 3.
Source: © Lu Guang/Greenpeace.

viewpoints. And, with a student cohort that is globally dispersed, it can be expected that students will bring to these forums diverse responses to imaging "the environment," thereby co-constructing global EC understandings. These kinds of collaborative tasks across geographical locations can provide a wide education in global environmental issues at the same time as teaching practical journalism skills.

By allowing students to progress from simple objectivist learning to constructive, collaborative, globally situated learning, this exercise demonstrates how the distance/online medium can fully accommodate the active, self-directed, and experiential study necessary for learning a subject like EJ well.

A good fit for teaching EJ

This discussion has shown that, if an EJ course is thoughtfully designed in terms of instructional strategy, content, structure, and technology, the non-university distance study medium can be an effective environment for teaching this aspect of EC. Online/distance EJ students who have actively and collaboratively constructed their own knowledge (rather than passively assimilating it) can become environmental communicators skilled in lateral thinking, problem-solving and teamwork, thereby broadening the real-world application possibilities for their EC learning. Studying EJ as part of a global community of learning—unique to online study—also offers students a truly global understanding of environmental issues and an environmental communicator's interaction with them.

There is also the fact that, in the digital media landscape, the journalist's role has drastically changed. Journalists no longer simply write or film or take pictures, or specialize in any particular subject, or part of the world. They do all of this and much more. No longer one-way transmitters of knowledge, they are curators, miners, verifiers, and *collaborators* (with the group/s once known as their public/s) in knowledge *construction*. If journalists are now in the business of collaborative knowledge construction with their digital publics, then surely it makes sense for journalism students to learn their profession collaboratively and constructively in an online community.

This is not to say there are no caveats regarding teaching EJ by distance. A major factor in distance/online learning is high drop-out rates. Studies have indicated these can be 25–40 percent, as opposed to 10–20 percent for on-campus students (Levy 2013, 186). An online course in any aspect of EC therefore risks losing more potential environmental communicators than a face-to-face version. With the aim of retaining as many future environmental communicators as initially sign up, course designers should carefully follow students' progress through the study materials to deduce points at which they tend to abandon a course. Fully online course material (as opposed to the printed course book of the EJ course discussed here) can offer useful digital feedback for tracking study pathways and actual coverage of learning material.

On a practical level, though online/distance EJ courses can have a degree of internationalism built in, it is difficult to gage in advance where exactly students might be studying from in order to make the course true for them. A successful EJ course would therefore need to prioritize co-construction of content through some kinds of collaborative projects to reflect the disparate environmental understandings of an international cohort. Online EJ courses must also point out that conditions for doing EJ are not the same worldwide—and, in some national contexts, EJ is even a dangerous profession. Depending on where they are located, students may therefore bring very varied, and illuminating, perspectives on EJ to their studies.

On a pedagogical level, while online EJ courses can implement a constructivist learning approach just as effectively as face-to-face courses, when the material being covered is fact-based or procedure orientated, the transmission/objectivist style may be more useful. In this way, basic information can quickly be imparted as background for more creative, applied problem-solving. However, as the discussion in this chapter has pointed out, it is not only *information* that the objectivist model of teaching conveys, but also ideologies, values, and constructions. The question remains, then, as to how best EJ instructors might educate environmental communicators who advocate for the planet, without too great a degree of transmission-style indoctrination in their teaching?

The online/distance medium and the non-university environment, then, seem to offer great opportunities for educating people with an interest in EJ to become ethical, globally minded environmental communicators. Although caveats and questions remain, these may be addressed as more distance courses in EJ are taught, and more EC/EJ instructors can reflexively examine their pedagogical practice across a range of learning contexts.

References

Anderson, A. (1997) *Media, culture and environment*, Routledge, Abingdon.

Arbaugh, J. and Rau, B. (2007) "A study of disciplinary, structural and behavioral effects on course outcomes in online MBA classes," *Decision Sciences Journal of Innovative Education*, 5(1): 65–95.

Bourdieu, P. (1991) *Language and symbolic power*, Polity, Cambridge.

Cantrill, J. and Oravec, C., eds. (1996) *The symbolic earth: Discourse and our creation of the environment*, University of Kentucky Press, Lexington.

Castells, M. (1996) *The rise of the network society: The information age: Economy, society and culture*, Blackwell, Oxford.

Chen, S. (2007) "Instructional design strategies for intensive online courses: An objectivist–constructivist blended approach," *Journal of Interactive Online Learning*, 6(1): 72–86.

Cottle, S. (2006) *Mediatized conflict: Developments in media and conflict studies*, Open University Press, Maidenhead.

Cox, R. (2007) "Nature's "crisis disciplines": Does environmental communication have an ethical duty?," *Environmental Communication*, 1(1): 5–20.

Cox, R. (2012) *Environmental communication and the public sphere* (3rd edn.), Sage Publications, Thousand Oaks.

Daley, J. (2010) "Why the Decline and Rebirth of Environmental Journalism Matters," Yale Climate Connections. Available at: www.yaleclimateconnections.org/2010/01/why-decline-rebirth-of-environmental-journalism-matters/ (accessed February 9, 2015).

Franzen, A. and Vogl, D. (2013) "Two decades of measuring environmental attitudes: A comparative analysis of 33 countries," *Global Environmental Change*, 23: 1001–8.

Hansen, A. (1991) "The media and the social construction of the environment," *Media, Culture and Society*, 13: 443–58.

Hansen, A. and Cox, R., eds. (2015) *The Routledge handbook of environment and communication*, Routledge, Abingdon.

Hiltz, S. and Goldman, R., eds. (2005) *Learning together online: Research on asynchronous learning networks*, Lawrence Erlbaum, Mahwah.

Jaffe, R., Moir, E., Swanson, E. and Wheeler, G. (2006) "E-mentoring for student success: Online mentoring and professional development for new science teachers," in Dede, C., ed., *Online professional development for teachers: Emerging models and methods*, Harvard Education Press, Cambridge, MA, pp. 89–116.

Jonassen, D. (1992) "Objectivism versus constructivism: Do we need a new philosophical paradigm?," *Educational Technology Research and Development*, 39(3): 5–14.

Kuhn, T. (1996) *The structure of scientific revolutions*, University of Chicago Press, Chicago.

Lakoff, G. (1987) *Women, fire, and dangerous things: What categories reveal about the mind*, University of Chicago Press, Chicago.

Lester, L. (2010) *Media and environment: Conflict, politics and the news*, Polity, Cambridge.

Levy, Y. (2013) "Comparing dropouts and persistence in e-learning courses," *Computers & Education*, 48: 185–204.

Mangen, A., Bente, R. and Kolbjørn, B. (2013) "Reading linear texts on paper versus computer screen: Effects on reading comprehension," *International Journal of Educational Research*, 58: 61–8.

Means, B., Toyama, Y., Murphy, R., Bakia, M. and Jones, K. (2009) *Evaluation of evidence-based practices in online learning: A meta-analysis and review of online learning studies*, US Department of Education Center for Technology in Learning. Available at: http://files.eric.ed.gov/fulltext/ED505824.pdf (accessed July 18, 2015).

Mizrachi, D. (2014) "Online or Print, which do students prefer?," in Kurbanoğlu, S., Špiranec, S., Grassian, E., Mizrachi, D. and Catts, R., eds., *Information literacy, lifelong learning and digital citizenship in the 21st century*, Springer International, Cham, Switzerland, pp. 733–42.

Moore, J., Dickson-Deane, C. and Galyen, K. (2011) "e-Learning, online learning, and distance learning environments: Are they the same?," *The Internet and Higher Education*, 14(2): 129–35.

Moore, M. and Kearsley, G. (2011) *Distance education: A systems view of online learning*, Wadsworth, Belmont.

Newman, P. and Fernandes, R. (2015) "A re-assessment of factors associated with environmental concern and behavior using the 2010 General Social Survey," *Environmental Education Research*, 22(2): 1–23.

Palocsay, S. and Stevens, S. (2008) "The study of effectiveness of web-based homework in teaching undergraduate business statistics," *Decision Sciences Journal of Innovative Education*, 6(2): 213–32.

Piaget, J. (1970) *Genetic epistemology*, Columbia University Press, New York.

PIEC (2015) *Trends in Visibility of Environmental Issues in the U.S. News Media*, Project for Improved Environmental Coverage Report. Available at: http://greeningthemedia.org/wp-content/uploads/PIEC-Trends-Report.pdf (accessed August 23, 2015).

Reilly, D., Stafford, R., Williams, K. and Corliss, S. (2014) "Evaluating the validity and applicability of automated essay scoring in two Massive Open Online Courses," *The International Review of Research in Open and Distance Learning*, 15(5): 83–98.

Sachsman, D., Simon, J. and Myer Valenti, J. (2008) "Environment reporters and US journalists: A comparative analysis," *Applied Environmental Education & Communication*, 7(1–2): 1–19.

Schell, P. and Janicki, T. (2012) "Online course pedagogy and the constructivist learning model," *Journal of the Southern Association for Information Systems*, 1(1): 26–36.

Taylor, C. (2002) "Teaching and learning online: The workers, the lurkers and the shirkers," *Distance Educator*. Available at: http://distance-educator.com/teaching-and-learning-online-the-workers-the-lurkers-and-the-shirkers/ (accessed July 15, 2015.

Thompson, J. (1995) *The media and modernity: A social theory of the media*, Polity, Cambridge.

Vrasidas, C. (2000) "Constructivism versus objectivism: Implications for interaction, course design, and evaluation in distance education," *International Journal of Educational Telecommunications*, 64: 339–62.

Vrasidas, C. and Glass, G. (2004) "Teacher professional development: Issues and trends," in Vrasidas, C. and Glass, G. V., eds., *Online professional development for teachers*, Information Age, Greenwich, pp. 1–12.

Vygotsky, L. (1978) *Mind in society*, Harvard University Press, Cambridge, MA.

Zhang, D. (2005) "Interactive multimedia-based e-learning: A study of effectiveness," *American Journal of Distance Education*, 19(3): 149–62.

Part III

Transformative practice

Nurturing change agents

This part focuses on transformative practice and transformational learning, and asks how we might engage a particular environmental communication ethics in pedagogy in and outside of the classroom.

11 Changing *our* environmental future

Student praxis through community inquiry

Eli Typhina[1]

> As I enter the classroom, I can see everyone is excited for the start of a new semester. After brief introductions, I ask students, "What environmental issues concern you?" Many hands go up and responses include water pollution and species extinction.
>
> "What are you going to do about these concerns?" I ask.
>
> After a moment, a student in the front row proclaims, "I'm going to reduce CO_2 by riding my bike to school every day."
>
> "Good idea," I offer, "but what about the other students who choose to continue driving their cars?" The room becomes silent, except for the squeak of chairs under the students' jittery legs.
>
> A student in the back of the room looks up from her iPad and sarcastically offers, "Just get the government to ban driving cars!"
>
> Turning his gaze from the window, a third student genuinely responds, "No … let's get some non-profits to run education campaigns and once people know the harm, they will stop driving their cars!"

This dialogue, one example of many I have had with students over the years, reveals students' perceptions of their place in environmental action: on the periphery. They know they can change their own behavior, but when it comes to the behavior of members of their community, they believe it is up to larger, more powerful organizations to solve environmental issues. What students do not see is the power they hold to ignite social change in their community through their passion, ideas, and creativity. To build student momentum, I created four strategies that I use in my upper division Environmental Communication course to help students gain the confidence, tools, and experience to engage their community in environmental action. The pedagogical approach grounding my teaching is that of Lave and Wenger's *communities of practice* (Smith 2003). Communities of practice are groups of people working toward a common goal; the language and artifacts developed in each group connect members to each other and impact members' orientation to other groups and to the world around them (Eckert 2006).

Using communities of practice in my environmental communication classroom helps students realize how to apply their knowledge of the world to solve environmental problems with their community. Students know a great deal about environmental problems, but they are unsure who will solve these problems and what their role would and should be in the solution. In this chapter I show how the environmental communication classroom can become a safe and creative space for students to transition from passive learners to active change agents through building relationships and collaborative partnerships with community organizations (Smith 2003). In my course, classroom readings and activities are paired with out-of-class projects students develop in collaboration with a community organization. The process of using classroom materials to develop ideas, resources, and relationships with their organization gives students the foundation to build competency as community change agents (Lave and Wenger 1991; Wenger 1998; Wenger 2000). The combined in- and out-of-class experience gives students the space and time to learn how to negotiate community needs and their own abilities: skills they can use throughout their lives (Wenger 2000).

This chapter outlines the curriculum I developed to help students connect their learning with the communities and organizations around them through application of environmental communication theories and perspectives. The environmental communication classroom is the perfect space to partner learning and doing because it provides the right amount of support, freedom, and, to a degree, mediation. Instructors serve an indispensable role in the development of students into change agents by providing class readings, activities, and support with outside projects. Both students and organizations approach collaborative projects with assumptions about the other's competencies; the instructor helps the two by translating between the groups and encouraging them to suspend their judgments to allow both parties to grow (Wenger 2000).

The initial challenge of a student working one-on-one with a community organization is the perception of power differences between the student and the organization. Both are familiar with the hierarchical power relationship of internships, where the organization assumes they need to tell the less knowledgeable student what to do. The communities of practice approach is just the opposite: it champions a horizontal power structure where the student is more knowledgeable in some domains than the organization and vice versa. I remind students that their experience with their organization should resemble a consultancy, not an internship. To establish and maintain this power dynamic, I encourage them to advise their organization of ways to become more environmentally friendly by referencing theories and scholars from class.

To foster an environment for student growth between classroom and community, I designed four strategies that guide my selection of readings, activities, and assignments (see Figure 11.1). I will briefly explain each strategy here, then each in more detail throughout the chapter. In the middle of the figure are *mini-activities*. These activities bring to life course readings by asking students to propose environmental interventions for businesses, non-profits, and government agencies. *Theory to action*, in the bottom left, acquaints students with environmental communication scholars and their perspectives through readings,

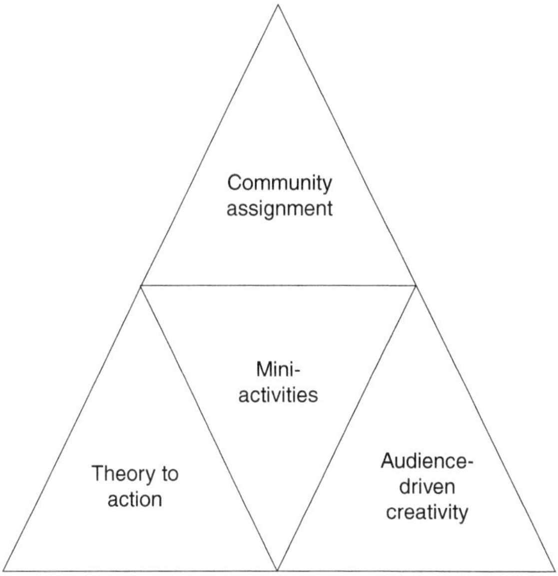

Figure 11.1 Four strategies for selection of readings, activities, and assignments.
Source: author.

which give students tools for crafting interventions in the mini-activities. *Community assignments* and *audience-driven creativity* completely immerse students in their local community through one-on-one relationship building.

In my environmental communication course, I draw from three theoretical perspectives. The first are rhetorical and critical perspectives, which help students see how communication propagates or subverts various forms of environmental behavior (Milstein 2009). One reading taking this perspective is Cox's (2013) *Environmental communication and the public sphere*, which familiarizes students with the many communities and viewpoints in environmental debates. The second perspective is behavioral, which helps students identify behavior change tactics through application of psychology and marketing theories. Books with this perspective include McKenzie-Mohr's (2011) *Fostering Sustainable Behavior* and Klöckner's (2015) *The Psychology of Pro-Environmental Communication*; both equip students with techniques and methods to intervene in the communities described in Cox's book. The third perspective, that of the student, I call the creative perspective because it comes about when students transform rhetorical and behavioral theories into creative problem-solving strategies for their community. There are no books explaining the 'creative' perspective because it is a perspective students surmise from their life experience, personality, learning in class, and interactions with their community of practice. The four strategies explained in this chapter aim to assist students to develop their creative perspective through activities that require them to play with their personal boundaries and to see how their unique character can contribute to solving their community's environmental problems. Through activities and community projects, students are able to suspend their identity as someone unable to create change and learn how to cross boundaries and connect with the diverse people within their community in such a way as to make change (Wenger 2000).

The remainder of the chapter describes the four strategies I use in my course to support students' transition from learners to community change agents. Each section is filled with example readings and activities illustrating how I encourage development of rhetorical, behavioral, and creative perspectives within student repertoire.

Mini-activities

Mini-activities are a hands-on way to introduce students to environmental communities of practice without leaving the classroom. These activities bring class readings to life by encouraging students to embody the personalities of characters in the readings through role-playing or simulating events from the readings. I bring props, such as costumes or food, and creative tools, such as poster paper, markers, or sidewalk chalk, which enable students to play with social norms and alternative endings in ways not possible outside the classroom. To foster a playful atmosphere, I offer crazy, outside the norm possibilities that get students' creative juices flowing. To increase student comfort in being outlandishly creative, I reassure them that what we do and discuss will not be graded or shared with outside audiences. The following two examples show how mini-activities help students to develop their *creative perspective*:

Activity: city council vote on Funpark Fyn's renovation

The objective of this activity is for students to learn to empathize and work with people holding environmental ideologies different from their own. The second chapter of Corbett (2006) accompanies this activity because it represents rhetorical perspectives with rich descriptions of environmental ideologies including unrestrained instrumentalism, conservation, preservation, values-driven ethics, and transformative ecologies.

The city council vote activity asks students to embody one of the five environmental ideologies in Corbett's (2006) second chapter and debate with groups of opposing ideologies how public land should be developed. I divide students into groups of three or four and ask them to draw a paper with one of the five ideologies from a bag. Then I present the following scenario:

> The city of Fyn has acquired the derelict amusement park property, Funpark Fyn, and the city council must vote on what to do with the property. Residential areas in Fyn have expanded, and the once remote

amusement park is now surrounded by houses; many residents are concerned with its safety. I show a clip of the park to give context. (Use Funpark Fyn in your internet search.) The council invited the five groups to the meeting to voice their concerns and provide suggestions for developing the property. The groups have approximately ten minutes to use their ideology to invent their organization's name, identify a plan for development, and write six arguments why the city council should implement their development plan.

Once all groups have their arguments prepared, I call the council meeting to order and ask that one group representing each ideology stand at the front of the room to present their idea for development. The remaining students and I stay seated and serve as city council members. The responsibility of the council is to use the rubric to decide which group will win our vote (and a bag of chocolates). The rubric states the winning group should do the following:

- present arguments that align with their ideology (i.e., if the group ideology is unrestrained instrumentalism, do they argue humans are the most important entity or do they argue against their ideology by saying all life possesses equal value?);
- adamantly engage other groups by asking them questions and challenging their ideologies; and
- express passion and interest in the outcome of the ruling (I encourage students to use gestural and vocal enthusiasm).

After students present their development plan, I open the floor for debate. Once all the groups have had a chance to engage in the debate I call time. We applaud the groups and the council members vote for the winning team. Class concludes with reflection of the activity on students' perceptions of environmental debates.

Activity: stop the grease monster

The purpose of this activity is for students to design an environmental behavior intervention to improve a human-caused environmental problem by integrating audience behavior and behavioral theories into the intervention's design. I use readings from McKenzie-Mohr (2011) and Fogg (2009) because their behavioral perspective offers ways to tweak components of an individual's environment to influence their motivation, ability, and social climate toward environmentally friendly behavior.

This mini-activity asks students to use components from McKenzie-Mohr and Fogg's readings to design a behavior intervention that will prevent community members from pouring food grease down the drain, which can cause sanitary sewer overflows and pollute waterways. I start by showing the call for solutions to the grease problem on our city's website, we watch video clips from YouTube, and then enact a kitchen scene where students dump food grease down a pretend drain. Students form groups and choose one of the behavior change tactics discussed in the readings to guide the design of their intervention. After drawing their intervention on a large poster board, the groups present their solution and we discuss how each group's solution connected to the audience's behavior and the components of their theory.

Community assignments

Students refine *creative perspectives* formed in the mini-activities into frameworks for change in community assignments. I design community assignments to serve as boundary objects between the practices of students' partner organizations and students' ideas for making these practices greener (Wenger 2000, 236). As students work with their organizations on assignments, they generate:

- *artifacts*: an informational video about the organization's environmental policies;
- discourses: shared ways of talking about the environmental concerns of the organization; and
- processes: an agreed upon environmental communication plan (Wenger 2000, 236).

Before the semester starts, I request students to research local government agencies, non-profits, or businesses they would like to work with during the semester, keeping in mind their partnership might turn into a job opportunity after the course. The organization does not need to be environmentally focused; however, they do need to be receptive to implementing environmental practices and spending time with the student. On the first day of class, I review the assignments and provide a letter students can present to perspective organizations. The letter includes assignment goals, deliverables for the organization, and the quantity of time organization staff will need to spend with the student for assignment completion. I emphasize that this is a partnership, not an internship. Students provide consultant-like services to the organization through a portfolio of materials that will improve the organization's environmental practices and community relationships. In return, organization staff provide their time and ideas during regularly scheduled meetings with the student. On the last day of class, I request that students bring their laptops and résumés so we can add their class experience to their résumé and to answer questions on how to use course experience to obtain career advancement. I also have students write thank you letters to their organization and inquire about the possibility of a paid or volunteer internship to continue gaining experience and building their resume.

Since the semester is fifteen weeks, I usually choose to design community assignments around one of two approaches:

- focus on theory and methods acquisition, where most of the semester is spent exploring theories in relationship to the organization and the last quarter of the semester involves brief data collection, analysis, and an intervention proposal; or
- focus on implementation, where the first quarter of the semester introduces a few theories and methods while the remainder of the semester has students collect data, analyze it, and implement their intervention with the organization.

The following example is a set of assignments following the theory and methods acquisition focus.

Assignment: blog portfolio

The blog portfolio is designed as a digital space for students to write down and work through their creative perspectives. It functions as a professional portfolio for the organization's reference and as a reference for future employers if students include it on their resume. At the completion of a set of rhetorical and behavioral readings in class, I ask students to meet with their organization and develop an intervention for environmentally friendly behavior based on the perspectives covered. I give students class time to meet with their organization because they are often too busy to meet with the organization during business hours outside of class. One-on-one meetings with the organization are the moments wherein students transform into change agents through development of their creative perspectives. For example, while discussing with an organization challenges to sustainability, students will propose a solution for the organization that draws from a perspective they learned in class that uniquely addresses the concerns of the organization. In that moment students realize that without their background and creativity, the solution could not have been found nor implemented. After their meetings, students draft blog posts describing their unique intervention (see next section for an example post).

Once I receive the posts, I provide suggestions on ways students can further integrate course concepts to solve the organization's problems, which they use to revise their post and bolster development of their creative perspectives. Toward the end of the semester, students choose one of their posts to develop. They collect data on their audience and develop a detailed description of their intervention so that the organization can implement it. At the end of the semester, I give students class time to present their portfolio and detailed intervention to their organization and discuss the possibility of gaining a paid internship (or volunteering) to implement their intervention with the organization. I evaluate student work based on their integration of traditional and behavioral perspectives, as well as development of their creative perspectives. The following are examples of organizations students worked with and the students' resulting portfolios:

Dry Cleaners
Post interventions proposed by Austin Long included using social media to share the green practices of a dry cleaning store with customers, incentivizing the organization's hanger recycling program, and painting the outside of the store with biodegradable paint to mimic and raise awareness of the store's green cleaning practices.
Blog portfolio: http://medlin-davis.weebly.com/

Apartment complex near a greenway trail
Sarah Drumheller proposed interventions in her posts such as promoting a green office for staff, designing a website to promote environmental communication with residents, and running a litter collection event near the apartment complex and greenway for staff and residents.
Blog portfolio: http://drumhellerenvironmental.weebly.com/

Theory to action

I often say to my students, "Stand on the shoulders of giants. When you present an idea to your organization, cite the scholar and theory that inspired it." When students cite readings from class, they feel more confident in proposing interventions to the organization and the organization feels more confident in implementing their recommendations. As the semester progresses, interactions with organizations continue to boost student confidence and their belief that they can be and are community change agents. Throughout the course, I offer students concepts from the rhetorical environmental communication perspective so they can gain new ways of approaching problems, while I offer behavioral perspectives to give students tools for measuring and influencing human behavior.

My aim in every class is to couple rhetorical and behavioral readings with mini-activities and community assignments. The following three examples are written in the order in which I teach them. I start by showing students how they can apply concepts through real-world examples, then I ask students to apply the concepts in groups through a mini-activity, and finally I ask them to apply the concepts on their own through their community assignment. I use the same readings and concepts for all three examples.

Example application of concepts

The following activities help students understand how to improve relationships with clients and employees through open dialogue about environmental issues. I use Kinsella (2004) because he argues for the importance of engaged, knowledgeable citizens who can effectively collaborate with technical experts in environmental decision-making. He points to the challenges of creating spaces where citizens and experts can equally share their unique knowledge to solve environmental problems. Kinsella's article provides a critique of communication practices and is focused on generating dialogue between parties. Because of this, I pair it with Kent and Taylor's (1998) and Reber and Kim's (2006) behavioral articles that describe *how* to create dialogic loops between parties in online environments for two-way flow of information as recommended by Kinsella.

Since there is no obvious connection between the articles, and the behavioral articles are not about environmental issues, I explicate in class how students can combine the results of both to solve community environmental problems. In doing this, I present four dialogic points, integrate Kinsella's perspectives, and provide examples from Patagonia's website. The points include the following:

- Create a digital dialogic loop by providing a space, such as an online forum or survey, where the public can ask questions and the organization can respond. This can be seen on Patagonia's website where they provide a chat window for customers to talk with them directly and immediately.
- Make the website useful to all publics by providing information on the organization's environmental ideologies, practices, and initiatives through an online press room and biographies of organization members. Patagonia does this through "The footprint chronicles" where they reveal the energy, shipping, CO_2, waste and other aspects of their production and supply chain to stakeholders.
- Present a beautiful, easy to use interface that can help stakeholders become knowledgeable and engaged in discourse through environmental awards and images of employees engaging in environmentally friendly practices. Patagonia displays badges, such as being a part of the 1 percent club, that show their environmental ideologies and mission quickly and easily.
- Generate return visits to promote healthy discussion by updating pertinent and valuable information through question and answer sessions with and public challenges for environmental action. An example of this from Patagonia is their Common Threads Partnership "take the pledge" challenge, where they ask their stakeholders to pledge to reduce waste by reselling, repairing, and reusing their products.

After reviewing the articles and the Patagonia website, I draw students' attention to a local issue for our mini-activity—horizontal drilling and hydraulic fracturing for oil and gas (commonly called fracking). Using their mobile device or laptop, students pull up our state's Department of Environment and Natural Resources website on the topic. Based on their exploration of this site, students discuss in pairs how a critical rhetorical scholar might interpret the website. They note the existence and/or lack of the four dialogic website components and, using Kent and Taylor and Reber and Kim's articles, they devise recommendations for improvement. Following this, we discuss what students found and recommend. The discussion is important for clearing any confusion students have as to how they can apply theories to real-world problems. There are no grades assigned to the activity in order to allow for open and creative discussion.

Building on the mini-activity, I ask students to apply what they learned in class to critique their organization's website for their community assignment. Students include in their post a description of stakeholder needs, how the website currently addresses those needs through dialogic means, and how the organization can improve dialogic communication through their website. As part of their course grade, students create a blog post of their comments and share it with their organization. I have found the time students spend on their own and with the organization crafting these posts are the most memorable and influential to their engagement in community action after the course.

Audience-driven creativity

The easiest way for students to join in a community of practice and form their own creative perspectives is through audience observation and engagement. Previously, I highlighted a few ways I encourage audience evaluation, such as providing opportunities for students to embody differing environmental ideologies, acting out scenarios, and talking with their organization. I also offer students materials that help them identify differences in audiences, such as Nielsen's (2016) PRIZM segmentation or Domingo et al.'s (2013) environmental segmentation. During our mini-activities, I explain to students how to observe and interview their audience to identify ideologies, barriers to behavior, and incentives to action. I also tell them their solution must be new, never done before. This often frustrates students, but it challenges them to move beyond interventions that are not appropriate for their audience to interventions that are unique to the students' personalities, the problem, and the organization. Before students embark on their final project, we use class time to conduct mock observations and interviews. The following mini-activity and community assignment show the importance of connecting the audience to students' interventions

Mini-activity

In the *stop the grease monster* mini-activity (mentioned earlier) students gain an understanding of their audience by first watching a video clip of a family engaging in kitchen grease disposal, and then they act out a scenario in class. For the scenario, I ask two students to stand at the front of the room wearing aprons and holding pans, and then I ask the remaining students to call out how the scenario should progress. As students start describing the scene, the acting students move about as if they are in a kitchen. I ask the actors and students to contribute commentary about how people in the scene might feel, what their actions are, and what barriers prevent them from enacting more environmentally friendly behavior. As the discussion unfolds, I write on the board what students identify as elements of their audience's behavior such as time, incentive, and ease of grease disposal. After reviewing the items on the board, I ask students to reflect on the theory we discussed in class and propose the type of communication we should use. Typically, they recommend placing a sign by the sink, and I respond by asking them how the sign helps their audience overcome the barriers we identified (i.e., time, incentive, and disposal location). After this, I break students into groups and ask them to devise a more effective form of environmental communication with their group, one that has not been tried before. The groups use markers and a large sheet of paper on the wall to design their solution, which they present at the end of class. Solutions have ranged from mobile apps that incentivize grease recycling to disposal jars that make a noise when they detect grease in the air.

Community assignment

About two-thirds of the way through the semester, I have students choose one environmental problem their organization wants to address and then spend the remainder of the semester exploring that problem. To connect students to their audience, I prepare a series of classes. The first class explores ways students can use theory to guide their observations and interviews; we also work on draft interview questions. In the following class, students bring their observation and interview questions; as a class we practice the interview questions students brought and, during our practice, I help them identify environmental ideologies, behaviors, and external factors that might influence their audience. I give students our third class as a workday to make their observations and interviews. In the fourth class, I pair students to help each other process findings and brainstorm interventions. The following are examples of student projects:

> *City Government*
> After observing and interviewing local pet owners, Katelan M. Hart identified that pet owners often did not dispose of their pets' waste because they did not know a location for disposal and they did not have much incentive. Using social norms and incentives as discussed in McKenzie-Mohr's (2011) book, she proposed a mobile application game, similar to the original Foursquare, where dog owners could see their location and the location of pet waste stations on a map. Every time a pet owner checked in at a station, they would receive redeemable points for prizes at local pet stores and veterinary offices, such as a dog treat or a percent off their next purchase.
>
> Project: http://environmentalcommunication.weebly.com/campaign-proposal-and-evaluation-method.html

> *Wildlife Non-Profit*
> Mackenzie Gokey wanted to help a local non-profit by informing and engaging community members in wildlife care and rehabilitation. After observing and interviewing community members engaging with wildlife at a science museum, she identified her audience as people in their twenties who most often used Instagram. Mackenzie applied components of a dialogic website (discussed earlier) to design an Instagram campaign that

allowed the organization and stakeholders to post pictures of wildlife they encountered. Both parties could comment on each other's images. The organization could also post questions to their followers and offer special, one-on-one interactions with wildlife for the most active follower (i.e., most retweets, likes, and personal photos shared).

Project: http://mackenziegokey.weebly.com/campaign–evaluation.html

What we can do with what we have

It's almost half way through the semester. I ask students, "Imagine your supervisor at work stops by your desk and tells you that you need to create some recycling signs and hang them up because everyone is throwing paper in the trash bin instead of the recycling bin. What would you say to your boss?"

Excitedly one student responds, "I would ask, 'How big do you want the poster?' "

I shake my head. "No."

Another student offers, "Visuals are important. I would say, 'What kind of pictures do you want?' "

Again, I shake my head. "No."

The student who always sits by the window retorts, "I would tell my boss, 'Signs are one possible solution. However, let me observe and interview employees to find out why they are not engaging in recycling behavior. Then, I will apply this knowledge to a theory I learned in my enviro com class, and propose a solution that will help employees overcome any barriers they have to recycling.' "

I include this dialogue as a reminder that change is not easy, but it is possible. Getting students to recognize that they are change agents and that they do not need to agree with what everyone tells them to do can be difficult. Additionally, encouraging an organization to see the value and knowledge students bring can also be challenging. I have found that taking a communities of practice approach is not always easy, but it is always rewarding. Through relationships with community organizations, students learn new ways of knowing and being in the world (Wegner 2000).

When I have coupled communities of practice with rhetorical, behavioral, and creative perspectives, I have seen the dynamic of my classroom, community, and students change. The *mini-activities* show students how they can become change agents in their community, while the *community assignments* give them experience in making community change. By exploring the ways rhetorical and behavioral perspectives can alter environmental outcomes in *theory to action* and *audience-driven creativity*, students have a foundation for developing their own creative perspectives. The examples, enactments, videos, and assignments bring to life all three perspectives by moving them from the pages of a book to the world of the student. Wegner (2000), founder of communities of practice, calls this way of knowing *identities* and argues that being able to engage and suspend one's identity allows an individual to learn ways of crossing boundaries through one's lived community experience. I think it is through the process of fun, creativity, and exploration that students truly learn many new identities and ways to be environmental change agents.

Note

1 I would like to thank the following students for allowing me to reference their projects in this book chapter: Sarah Drumheller, Mackenzie Gokey, Katelan M. Hart, and Austin Long.

References

Corbett, J. B. (2006) "A spectrum of environmental ideologies," in *Communicating nature: How we create and understand environmental messages*, Island Press, Washington, DC.

Cox, R. (2013) *Environmental communication and the public sphere* (3rd edn.), Sage Publications, Thousand Oaks.

Domingo, L., Rasamoelina, F., Bouchard, C., Brissaud, D. and Aoussat, A. (2013) "Using environmental segmentation to perform ecodesign with users." Presentation at International Conference on Engineering Design (ICED) August, Seoul, 198. Available at: https://hal.archives-ouvertes.fr/hal-00870534 (accessed October 20, 2016).

Eckert, P. (2006) "Communities of practice," *Encyclopedia of language and linguistics*, 2: 683–5.

Fogg, B. (2009) "A behavior model for persuasive design." Paper presented at the Proceedings of the 4th International Conference on Persuasive Technology, Claremont, California.

Kent, M. L. and Taylor, M. (1998) "Building dialogic relationships through the World Wide Web," *Public Relations Review*, 24(3): 321–34.

Kinsella, W. J. (2004) "Public expertise: A foundation for citizen participation in energy and environmental decisions," in Depoe, S., Delicath, J. and Elsenbeer, M.-F., eds., *Communication and public participation in environmental decision making*, State University of New York Press, Albany, pp. 83–95.

Klöckner, C. (2015) *The Psychology of Pro-Environmental Communication*, Palgrave Macmillan, London.

Lave, J. and Wenger E. (1991) *Situated learning: Legitimate peripheral participation*, Cambridge University Press, Cambridge.

McKenzie-Mohr, D. (2011) "Fostering sustainable behavior: Community-based social marketing." Available at: www.cbsm.com/pages/guide/preface/ (accessed September 29, 2016); see also www.cbsm.com/public/world.lasso (accessed October 19, 2016).

Milstein, T. (2009) "Environmental communication theories," in Littlejohn, S. W. and Foss, K. A, eds., *Encyclopedia of communication theory*, Sage Publications, Thousand Oaks, pp. 344–9.

Nielsen Company (2016) "Nielsen PRIZM." Available at: https://segmentationsolutions.nielsen.com/mybestsegments/Default.jsp?ID=70 (accessed October 20, 2016).

Reber, B. H. and Kim, J. K. (2006) "How activist groups use websites in media relations: Evaluating online press rooms," *Journal of Public Relations Research*, 18(4): 313–33.

Smith, M. K. (2003) "Jean Lave, Etienne Wenger and Communities of practice," *The encyclopedia of informal education*. Available at: www.infed.org/biblio/communities_of_practice.htm (accessed October 19, 2016).

Wenger, E. (1998) "Communities of practice: Learning as a social system," *Systems Thinker*, 9(5): 2–3.

Wenger, E. (2000) "Communities of practice and social learning systems," *Organization*, 7(2): 225–46.

12 Storytelling as action

Exploring transformation

Mairi Pileggi and Eric Morgan

Whether at a personal or a societal level, environmental communication teachers and practitioners seek to encourage transformative learning that fosters a sense of civic identity in their students and supports active community engagement around environmental issues. The crux of this process lies in shifting and transforming the identities of self and communities, which are symbiotically linked, through a shift in communication practices. One way to shift communication practices is to change the narratives that circulate within and among communities. More specifically, educators can help students tell a story about themselves as community members. As a universal form of communication, narratives have the potential to greatly affect the actions of human agents in the world. In this perspective, students are encouraged to cast themselves as change agents when they narrate their stories and chronicle their variations in time. This chapter provides an account of one approach to storytelling in the environmental communication classroom and, in doing so, illustrates a practice of transformative learning. More specifically, we demonstrate how using Marshall Ganz's (2007, 2011) work on public narrative gave one teacher and her students a way to make visible the journey to action.

The transformative power of storytelling

Storytelling's role as pedagogical practice is not new (see Connelly and Clandinin 1990; Deniston-Trotcha 2003 [1998]). Certainly in journalism and media studies storytelling is key. Likewise, in science communication, Dahlstrom (2014) notes that although narrative cannot substitute for scientific reasoning, storytelling may be an effective way to engage lay audiences in science. In environmental communication, storytelling and "restorying" (Boje and Massoud 2013) can play a critical role in the transformation of learners regarding environmental concerns. This resonates for many teachers of environmental communication among whom one goal is to develop critical thinking skills around the exigencies of environment. As Ganz (2011, 1) notes, "Stories not only teach us how to act—they *inspire* us to act."

Over the past half century, the study of narrative has been a key feature of communication scholarship. While there have been several different veins of narrative scholarship, perhaps the most well-known is Walter Fisher's (1987, 1989) narrative paradigm, which highlights ways in which humans use stories to comprehend and to navigate their lives. In direct contrast with other theories of persuasion based in rationality, Fisher postulates the world itself is comprised of stories available for selection and performance by individuals. As people engage these stories, they constantly recreate their lives, thus becoming change agents by changing their stories.

Another well-developed approach to the use of narrative lies in the perspective developed primarily by Barnett Pearce and Vernon Cronen (1980) known as the coordinated management of meaning (CMM). Briefly, scholars in this tradition argue stories help us to coordinate our interactions and thus to make sense of various contexts. Telling stories serves this purpose, as do the different types of story: stories lived, told, untold, unheard, unknown, and untellable. These types provide a framework to assess the coherence of an interaction. They also can be used to understand and explain identity transformation. For example, I (Morgan) frequently have students in environmental communication classes who tell stories grounded in the premise of an inability to "do anything." These stories have components that justify this position through reprioritizing different spheres of life. For example, a common student narrative is one in which the student is "poor" and "busy." While it would be nice to get involved, so the story goes, there just aren't enough hours in the day or resources. They are stories that are both told and lived, and there is a high degree of coherence between these two narrative perspectives. On this basis, the goal of transformative environmental communication pedagogy is to bring to light the stories that remain untold and the stories that remain possible. One way to engage transformation in the classroom is to engage the diversity of stories told, particularly those from non-Western cultures. Doing so sheds light on the structure of existing stories and highlights spaces for change.

Another perspective on the role of narrative relevant to environmental communication pedagogy is that of Michel de Certeau. In *The practice of everyday life*, de Certeau (2012 [1984]) offers a broader theoretical frame that accounts for how stories open "*spaces*" *of transformation*, a move that is particularly critical in the context of environmental communication. In environmental communication, stories help shape our relationships with the wider earth and with other humans; they help us make meaning in our world and offer a way to share and transmit culture. Our apprehension of the environment, according to de Certeau, occurs from a dual perspective: a *topical* one (where we are) and a *topological* one (how we go to where we need to be). As a consequence, our description of space fluctuates between the static and the dynamic. We see this movement in narrative, for example, between the "once upon a time" and the hero's journey. De Certeau points out that these descriptions are paramount in the definitions of boundaries, which consist "in narrative contracts and compilation of stories ... composed of fragments drawn from earlier stories and fitted together in a makeshift fashion (*bricolés*)" (ibid., 122). Storytelling, thus, becomes foundational in delineating limits, a frontier. A frontier in this sense can be considered that place between a lived story and the potential of other stories to be told, lived, and created. A frontier is not a "nowhere" on a map; it is a point of differentiation as well as a point of contact. It is an in-between, best fitted for agents of change to take on "a mediating role" (ibid., 127).

This analysis of physical space, which de Certeau (ibid., 125) describes through storytelling, also resonates in the abstract space of ideas where the frontiers are points of contention and where storytelling "opens a legitimate theater for practical actions ... [and] creates a field that authorizes ... social actions." As they demarcate boundaries, stories act as both "frontier" and "bridge." However contradictory this may seem, as the storyteller transcends boundaries through telling the story, he or she opens possibilities of experiencing other ways of being in the world. This transformative quality is a critical element in environmental communication, in which storytelling presents itself as a potentially useful and perhaps underused pedagogical tool to transcend boundaries.

The perspectives offered here establish the transformative possibilities that arise from using narrative as a pedagogical tool. In his work on public narrative, Ganz (2007) offers one way of using storytelling to implement change, and thus actualize the transformative power of narrative. He identifies three stories that comprise public narrative:

- a "story of self" focuses on one's calling;
- a "story of us" focuses on the calling of the community; and
- a "story of now" focuses on the challenges and possibilities facing communities.

His work has been adapted to grassroots organizing, from local immigration rights groups to the Obama presidential campaigns to helping environmental organizations craft messages around issues such as climate change and disruption (See climateaccess.org and 350.org). As the following extended example illustrates, his work is an effective approach to environmental communication pedagogy.

Storytelling as action

Fall semester 2014 at Dominican University of California is dry. California rainfall has reached record lows; reservoirs are at 50 percent of capacity. Governor Brown has called for conservation measures, asking citizens to cut consumption by 25 percent. My (Pileggi) school, a liberal arts and professional university with a student body of about 2,000, is known locally for its beautifully landscaped grounds. Although the university is moving toward xeriscaping, planting native and drought tolerant plants to reduce water use, lush green lawns carpet the landscape. As a student and I walk to class together one warm September afternoon, the sprinklers erupt. Startled, the student turns to me, and remarks, "I thought we were in a drought?" This incident, the incongruity of watering green lawns during a drought, exemplifies a disorienting dilemma for my student, and marks a first step in her transformative learning. A disorienting dilemma, as

Cranton (2002, 66) explains, is "an activating event that typically exposes a discrepancy between what a person has always assumed to be true and what has just been experienced, heard or read."

As individuals and as a community at Dominican University of CA, we strive to embody civic responsibility and to uphold a commitment to sustainability. We identify ourselves as both responsible and caring citizens. These are our foundational stories, the stories we try to live by, but they are not our only stories. The news stories describing the state of California's drought, the University story explaining its environmental policies, and the student's story, albeit short, expressing shock at the apparent dissonance of the situation are examples of competing or contradictory stories that circulate within the University community. This incident, which exemplifies environmental concern expressed through storytelling, became a catalyst for change in our environmental communication class.

Public narrative

Adapting Ganz's (2007) work on public narrative to an educational context helps students begin to reflect on their values and to understand how these values shape the stories students share about themselves, their "story of self." Moving from the personal to the collective, constructing a "story of us" helps the class shape the narrative of their environmental communication learning community. Finally, as the class focuses on a specific environmental problem, constructing a "story of now" moves the learning community to action. Although important in the classroom, the process of creating a public narrative also models skills that are valuable in organizations students may work with in the future. Thus, student practice has relevance beyond the educational setting. What follows is an account of how using Ganz's (2011) three stories worked in an environmental communication class.

The story of self

Students read Ganz's (2011) "Public narrative," which clearly explains the importance of storytelling, and his worksheet, "Telling your public story" (Ganz 2007), which gives direction for the assignment. As Ganz underscores, the work of articulating public narrative is a "leadership art" that offers a way to "translate values into action: engaging heart, head, and hands" (Ganz 2011, 1). He is adamant that a key step is articulating and understanding a "series of choice points," those challenges, choices, and outcomes that map the story of one's life (ibid.). The first story in the assignment, the story of self, gives students an opportunity to learn more about each other and to begin to build community. This activity is also useful in organizations and becomes a tool students can use in a work situation to shape organizational culture.

Before we began the story of self exercise, I invited a colleague from Career Services to lead a workshop with the class. First, we used media examples to prompt dialogue to understand what we mean by values. Car commercials were helpful for this exercise as they are short, often familiar to students, and promote specific values. Once it was clear from our discussion the class understood what was meant by values, students then worked individually with the Knowdell™ Career Values cards and worksheet (an Internet search will point to other useful ways to clarify values).

We began the next class with a timed writing exercise where students were asked to reflect on the values they identified as important. We continued our discussion of values and how they help to shape our lives. From this discussion we moved to an exploration of why we tell stories and reviewed elements of a story such as setting, character, plot, and moral. We then assessed aspects of storytelling in terms of Ganz's structure of challenge, choice, and outcome.

These exercises prepared students to complete the story of self. Working at home, students used the "Worksheet: Telling your public story," which helps focus on "the milestones and experiences that have brought you to this moment" (Ganz 2011, 2). They were asked to reflect on how specific challenges and choices helped them become who they are today. In this reflection, they were also reminded to include place. They then shared their stories in class. Small group discussions focused on values and emotions. Prompts such as "What did you feel?" "Why?" and "What values did you hear?" helped facilitate dialogue. As groups reported out to the full class, some students referenced specific aspects of their lives. One student found a sense of peace in her mother's garden and a connection to the environment; another felt her "love of animals" helped her learn compassion. One student wrote, "I lived in West Palm Beach, Florida for the first six years of my life … The smell of sea breeze that wafted in from the shore is deeply rooted in my memories of home." We reflected on the values manifest in specific stories, and through class discussion, the following group values emerged: hope, faith, compassion, and family/community. This exercise began to unveil the ground of a student's positionality. That is, in telling their stories students could begin to understand how their values are shaped in part by their experiences. Hearing the stories of their classmates, they also could come to see that the values shaping their perspectives may be shared by others or may be very different.

The story of us

The stories we tell about ourselves help to craft a larger social story. Creating the story of us, Ganz (2011) explains, articulates the values that motivate a community. Moreover, as we craft this story, we are, through our critical reflection, articulating those important values that give structure to our identities as distinct from others. The story of us asked students to consider who they were as a class, to clarify their values and experiences, and to reflect deeply on "why they do what they do" (Ganz 2007, 2).

Following an in-class brainstorming session, students decided "relationship" was a key value. Lively discussion ensued as they examined their relationships within the university, identifying roommates, teammates, classmates, teachers, staff, the physical campus, and the Earth. They decided "relationship" could translate as "community," an important institutional value reflected in their university's pillars or values. Dominican University of CA identifies four values: community, service, reflection, and study. These values stem from the university's history (Dominican Sisters of San Rafael founded the college) and are reflected in the university's mission, which also includes preparing responsible leaders and a commitment to sustainability. We focused on what community means for the class and discussed who the community is that shares our values. Students identified The Green Club, Campus Ministry, Pride Club, conservation biology class, groundskeepers, facilities, administration, and specific faculty members. At this point, students began to create their "vision" for the university. Everyone agreed Dominican University of CA is known for its beautiful campus. The vision then was that the University would shift to a beautiful campus that respects the environment, from toilets to lawns.

From an environmental communication perspective, this exercise teaches students an important step in designing a campaign. They were able to clearly articulate personal and group values, which could serve as a common ground for their media campaign. Practically, this exercise helped students make visible their personal networks, define their audience, understand who they might collaborate with on campus, and consider how they would share their story more broadly through various media platforms.

The story of now

The story of now moves the narrative to action, asking, "What do we need to change? What can I do about it?" In class discussions, students generated answers to questions focused on the challenges and choices the community faced. After thoughtful discussion, they were asked to begin brainstorming specific actions they might take. At this point, the student who witnessed the sprinklers on the lawns shared her experience. She referenced the drought and the call to reduce water usage. The group talked about the University's responsibilities, and the class agreed the University should honor its mission to prepare socially responsible citizens and to act sustainably. We talked about what that meant for us. One student said we should ask students to commit to reducing water, and dialogue shifted to brainstorming ideas and solutions. As we talked about the importance of individual actions, our conversation shifted again to larger questions and root causes—for example, consumerism, an ideology that doesn't allow for time to reflect, and social inequity were key. The students then crafted the following story of now: Dominican University of CA should take the lead in educating students and the community about the drought and should institute policies and practices that use water responsibly.

Taking action

Students recognized they didn't know enough about the drought to talk about it. They also were concerned others might be as uneducated as they were. As a first step they began to identify what they should know, such as where our water originated, how it was distributed, and where it would end up. To find out, a group met with officials from the local water district. Others wanted to understand what we mean by a drought and what effects it had. They contacted scientists and local and state government organizations to learn more. Using Survey Monkey, they polled other students to find out what they knew and were surprised to learn most students had not known California was in a drought. They then shared their knowledge, producing a radio show, a short film, articles for the student newspaper, and a letter to the editor of a local newspaper. They also held an event at the student cafeteria where they had a booth explaining the drought in California and asked students, faculty, and staff to pledge to conserve water. Students gave examples of how one might do this. More than 120 people signed. The Green Club, which partnered with the class, had a membership drive during the event and 10 students signed on.

At our last class, students reflected on their work. They noted that while it was important to raise awareness, there was no way of knowing whether those who pledged would change their behaviors. Some wondered how long they would continue their conservation practices once the class had ended. They agreed that, while personal action that changed individual behavior was important, there needed to be a systematic way to effect structural changes, such as policies. Several students reminded the class that, early in our process, they wanted to approach the University president and demand immediate reductions in water use. Our failure to engage directly with University decision makers, other students noted, was a barrier to effecting structural change. For future actions, we agreed we had to clearly identify key stakeholders and include them in conversations from the beginning.

Conclusion

In this class, using Ganz's (2007, 2011) framework for public narrative became a transformative experience for students. It helped students move from being "passive learners who accept the definitions (or situations) provided by others to become active learners able to define meanings and gain new perspectives on their own" (Mezirow 1991, 151). Critical self-refection, motivated by storytelling, addresses aspects of personal learning through the story of self. Collaborating to construct a story of us sparks reflection and learning at the institutional level. Finally, moving to the story of now moves students to action, the goal of transformative learning.

More specifically, students were able to identify and find shared values, a key component to communicating about the environment. They were able to design and employ a range of messages across media to educate their community and to spark dialogue about an important environmental problem. And they were able to reflect on ways to improve the effectiveness of their activism by considering ways they could better bring about structural change in the future. Thus, using storytelling in the environmental communication classroom could be an important pedagogical tool in a process of shaping students' emergent identities as environmental activists. Finally, reflecting on our pedagogy, we realize attention to assessment is key for future classes to document the degree to which students are developing an activist identity and to demonstrate ways in which the curriculum specifically fosters this.

References

Boje, D. M. and Massoud, J. A. (2013) "Determining environmental values: storytelling at BP," in Appleton, J., ed., *Values in Sustainable Development*, Routledge, London, pp. 68–77.

de Certeau, M. (2012 [1984]) *The Practice of Everyday Life*, trans. Steven Rendall, University of California Press, Berkeley.

Connelly, F. M. and Clandinin, D. J. (1990) "Stories of experience and narrative inquiry," *Educational Researcher*, 19: 2–14.

Cranton, P. (2002) "Teaching for transformation," *New directions for adult and continuing education*, 93: 63–71.

Dahlstrom, M. F. (2014) "Using narratives and storytelling to communicate science with nonexpert audiences," *Proceedings of the National Academy of Sciences*, 111(4): 13614–20. Available at: www.pnas.org/content/111/Supplement_4/13614.full (accessed July 18, 2016).

Deniston-Trochta, G. (2003 [1998]) "The meaning of storytelling as pedagogy," *Visual Arts Research*, 29: 103–8.

Fisher, W. R. (1987) *Human communication as a narration: Toward a philosophy of reason, value, and action*, University of South Carolina Press, Columbia.

Fisher, W. (1989) "Clarifying the narrative paradigm," *Communication Monographs*, 56: 55–8.

Ganz, M. (2007) "Worksheet: Telling your public story: Self, us, now." Available at: www.welcomingrefugees.org/sites/default/files/documents/resources/Public%20Story%20Worksheet07Ganz.pdf (accessed October 20, 2016).

Ganz, M. (2011) "Public narrative, collective action, and power," in Odugbemi, S. and Lee, T., eds., *Accountability through public opinion: From inertia to public action*, World Bank e-Library, pp. 273–90. Available at: http://marshallganz.usmblogs.com/files/2012/08/Public-Narrative-Collective-Action-and-Power.pdf (accessed April 1, 2015).

Mezirow, J. (1991) *Transformative Dimensions of Adult Learning*, Jossey-Bass, San Francisco.

Pearce, W. B. and Cronen, V. (1980) *Communication, action, and meaning: The creation of social realities*, Praeger, New York.

13 *Insider Windows* in Nepal

A critical pedagogy for environmental communication

Grady Walker

This chapter explores ways in which participants engaging in a critical pedagogy praxis in Nepal generated subjective learning themes related to their environment and society. Generative critical education stands in contrast to top-down methods of formal teaching and learning. To begin the discussion, let us consider two of the functions of environmental communication: to be pragmatic and to be constitutive (Cox 2013, 19). As human society deepens its understanding of the implications of life in the Anthropocene, the current geological age in which humans are the dominant influence on the environment, the need for strategic pragmatic environmental communication has become increasingly evident. Whether in the form of public awareness campaigns, direct action and protest, or consumer education, many diverse institutions are arising to answer the call. Among these is the university, charged with developing the future stewards of our planet. Ironically, the pragmatic nature of environmental communication messages has also made them vulnerable to cooption and dilution in the form of greenwashing, manufactured uncertainty, and a deceptive corporate environmental and social responsibility agenda. It is out of the matrix of higher education and into a world of competing ecological problem/solution narratives that today's graduates emerge. Will the environmental crisis we are facing be solved by market-based mechanisms, technological innovation, or blunt regulations? Unfortunately, this question is indicative of an understanding of ecological justice that does not interrogate the system as a whole.

Scaffolding the pragmatic function of environmental communication is its constitutive function. Environmentalism, broadly, is based on certain value-based understandings of the natural world. What we consider our environmental worldview has been articulated, or is implicitly understood, based on value constructions that use the building materials of constitutive environmental communication. Therefore, during the course of our lives, we compose understandings of the natural world that we use as reference points to initially validate or reject pragmatic environmental communication.

In the following sections I will discuss the theoretical foundations of a critical pedagogy field study based in Nepal and provide a summary of some of the outcomes insofar as they relate to constitutive environmental communication. By drawing on theory and the experiences of the participants in this study, this chapter will illustrate how informal critical learning can constitute subjective environmental values among learners based on justice, confirmed through action, and resulting in interventions in reality that are directed toward a better world, which is the goal of critical pedagogy.

Background

Critical pedagogy, critical theory, and ecopedagogy

Without departing too far from the subject of this chapter, it may be helpful for readers unfamiliar with critical pedagogy and its theoretical foundations to gain a better understanding of some of its assumptions. The goal of critical pedagogy has always been to move beyond the description and analysis of the world and toward its transformation by challenging the subject/object narrative Western culture has historically applied to both society and nature. Critical pedagogy holds the position that the current dominant socio-economic system is oppressive, and true social change must be systemic (Giroux 2011; Macrine et al. 2010). Major contributions to the field began with philosopher and educator John Dewey, who believed in the centrality of thinking and reflection in education, and that free interaction with one's environment is an essential part of knowledge construction (Darder et al. 2009, 3). Ivan Illich (1970) also made significant early contributions. Through his work among the poor in New York City in the 1950s and 1960s, Illich sought not just to provide criticism of growth-based economic rationality and the society that endorses it, but also to advance his vision of an alternative to what he considered a crisis of politics and unfettered technology (Darder et al. 2009, 5). Illich, among others, is credited with inspiring Paulo Freire, who is considered the preeminent theorist behind contemporary critical pedagogy. Freire understood that critical pedagogy would have to adapt to different contexts and to the dynamic nature of oppressive systems.

Indeed, the *raison d'être* of critical pedagogy is to endeavor toward justice within the human condition; but, based on its positioning within the wider landscape of critical theory, there is an understanding that social justice and environmental justice are inseparable insofar as both are reliant on systemic change. Ecopedagogical frames emerge from this understanding. All critical pedagogical thought draws heavily from critical theory and is bound to the work of the Frankfurt School, and, among them, Herbert Marcuse. Marcuse (1972, 61) argues that the domination of humans and nature are linked; both are based on the same rationality, and the logic of humans dominating humans can be overcome by challenging its roots in the social and scientific exploitation of nature. He is clear when placing blame: "It is obvious to what extent the violation of nature is inseparable from the economy of capitalism" (ibid.). The establishment, Marcuse asserts, deprives humans, who are being forced to waste their lives in "unending competitive performances," the opportunity of finding themselves in nature, which possesses life-enhancing forces. It prevents them "recognizing nature as *subject* in its own right—a subject with which to live in a common universe" (ibid., 60).

It is not a surprise, then, that ecopedagogy scholars argue in favor of Marcusian notions of pedagogy when tasked with developing a theory of radical education directed toward the ecological struggle (Kahn 2005). Insofar as social and ecological justice are inextricably linked, ecopedagogy attempts to develop a "moral character to aid in the survival of human societies and … Earth as a whole" (Milstein 2012, 4). While Freirian critical pedagogy has been perceived as falling short on addressing environmental issues, ecopedagogy carries certain a priori assumptions that direct pedagogy toward a more just world both socially and ecologically. This is indeed in harmony with Freire's overarching vision of critical pedagogy, which, he emphasizes, must be directive, and should always remain open to reinvention (Darder et al. 2009, 16; Freire and Shor 1987, 22–3). What the brief review of this study will show is that, even without the a priori assumptions specific to ecopedagogy, a critical pedagogy based on subject-generated learning content can still function as constitutive environmental communication.

Praxis, conscientization, and themes

Critical theory is not only a school of thought but also a "process of critique" (Giroux 2009, 27). This process allows for the uncovering of alternatives to constructions of society that hide behind taken-for-granted norms. These hidden constructions can be called *blocked subjectivities*, and, indeed, it is the purpose of *praxis* in critical pedagogy to unblock them.

The key element in the *Insider Windows* project, which I developed as part of a doctoral research project undertaken in Nepal, was its praxis. In the context of critical pedagogy, the meaning of praxis goes beyond understanding it as mere practice, distinct from theory. According to Freire (1970, 79), praxis is "the action and reflection of men and women upon their world in order to transform it." Au (2007) writes that praxis is central to critical pedagogy. He highlights its position within the overarching critical pedagogical activity, explaining that it gives participants the opportunity to reflect critically and subjectively on their reality and take transformative action to change that reality for the better, while deepening consciousness in the process.

Freire chose the Portuguese term *conscientização* to represent this deepening of consciousness. Now widely known as conscientization, it is defined in its simplest form as "the process by which students, as empowered subjects, achieve a deepening awareness of the social realities which shape their lives and discover their own capacities to recreate them" (Darder et al. 2009, 14).

The praxis I developed for this study to foster conscientization is centered on two pedagogical themes laid out by Shor (1992)—topical and generative:[1]

- A *topical theme* is an issue worthy of critical examination but is not necessarily generated through consultation. It may or may not be something that directly affects the local community, and is, in general, selected by an educator with due consideration of context and relevance.

- *Generative themes* are based on the experiences and understanding of participants, and represent their reality in context. The critical educator, in collaboration and negotiation with participants, helps these themes emerge through their shared understanding of the context. Pedagogy through the use of generative themes engenders a constitutive form of learning. Participants articulate their own understandings of issues that directly impact them, and, through consultation, combine individual insights to form a coherent understanding shared by the members of the group. This new understanding was previously hidden. It is through the use of generative themes that environmental issues impacting a community can be shared and verified. In other words, understandings about the environment and its degradation that were previously abstract can be reified through group analysis and praxis and related back to the lives of individual members of a group, or the community as a whole.

In the sections that follow, I will discuss the way in which the use of generative themes in pedagogy can be a powerful source of constitutive environmental communication.

Operationalizing conscientization: a framework

Conscientization occurs through the interplay of two types of knowledge (Morrow and Torres 2002, 46). The first, for the purposes of this study, will be called *outsider knowledge*, and the second, *insider knowledge*. I adapted the Johari Window, an analytical technique developed by psychologists in the 1950s, to provide a framework for visualizing the different types of knowledge and how they are related to the above themes (see Table 13.1).

Outsider knowledge, when introduced to an educational program, equates to expert knowledge and academic theory. Examples of this might include the science behind the anthropogenic climate crisis and the theories supporting a carbon-trading market. This knowledge is situated behind Window 2 of the Johari Window (Table 13.1). By contrast, insider knowledge refers to indigenous or local knowledge, and this is positioned behind Window 3. Using the

Table 13.1 Johari window: outsider and insider knowledge

	Insider knows	*Insider does not know*
Outsider knows	**Window 1: Open Knowledge** Empirical knowledge shared by insiders and outsiders *Topical themes*	**Window 2: Outsider Knowledge** Expert/intellectual knowledge owned by outsiders *Topical and Academic Themes*
Outsider does not know	**Window 3: Insider Knowledge** Local and indigenous knowledge *Generative themes*	**Window 4: Unknown/Blocked subjectivity** Revealed through conscientization

same example above, this could be seasonal knowledge about the climate that local farmers possess—frequency of rains, when to plant, etc. Behind Window 1 is the shared open knowledge: the climate is changing in the region. Both insiders and outsiders in the context above know this, but construct their knowledge differently. When this framework is applied to voices of the oppressed, Window 3 has been neglected and can be considered metaphorically opaque; it has shrunk and the voices behind it have been devalued. The insiders then devalue their *own* voices. A self-rejection of voice results in what Freire (1970, 30), drawing much from Erich Fromm, calls *the culture of silence*: the outcome of economic, political, and paternalistic social domination, all buttressed by the structures of mainstream education.

According to Morrow and Torres (2002, 102), members of a culture of silence can use dialogical learning to transform their accumulated experience into critical knowledge and understanding. Therefore, the directive elements of critical pedagogy have a final goal: *for participants to emerge from the culture of silence by way of conscientization*. This is realized by enlarging Window 3 and shrinking Window 4 through the constitution of new knowledge, which, in Freire's perspective, should be "based on the experience of learners in revising their own self-understanding" (ibid., 46). This approach precludes, to an extent, drawing knowledge into Window 3 from Windows 1 and 2; that is the dominant narrative. The eventual enlargement of Window 1 via Window 3 represents a de-privileging of objective outsider knowledge and a more balanced reimagining of what constitutes open knowledge.

Because these windows represent a framework level accounting of different kinds of knowledge, they can be applied to any type of communication involving insider and outsider dialectic. When applied to the example of environmental communication, this pathway from hidden knowledge to open knowledge enables the uncovering of subjective environmental value formations that may have been blocked. Consumerism and materialism, for example, when understood as outsider values injected into many indigenous communities around the world, may have replaced local values. These subjectivities, when unblocked, may offer alternatives to harmful outsider values and can contribute to a richer source of environmental understanding behind the window of open knowledge.

Insider Windows

In the *Insider Windows* project in 2014, youth participants in Nepal made short movies to investigate learning themes they generated through group consultation. Two young Nepalese women, whom I hired as critical educators and trained over a period of three months, recruited other young adults to participate in a video production class. Having grown up in Nepal myself, I speak the language and have knowledge of the culture; I knew, therefore, that hiring young men would mean they would be more likely to dominate the group consultations. As a non-Nepali I was a clear outsider, so I designed the research in a way that would limit the observer effect. I never engaged with the participants directly but only observed the video production classes unobtrusively with my research assistant.

The opportunity to learn basic video production was what initially attracted the participants to the project. They ranged in age from 18 to 30 and were predominantly young women. The movies they produced in small subgroups of three or four members were presented to the entire class for their analysis of generated themes. Cameras with built-in projectors made it possible for participants to screen their films in the classroom, even when there was no electricity. Environmental issues were commonly represented in the movies made during the praxis, not because the classes directed participants toward those issues, but because participants recognized their importance in the community when consulting together about a generative theme. By articulating an insider understanding of these environmental issues, the educators and participants not only shared environmental messages horizontally and vertically in their communities through organized and ad hoc film screenings but also decoded their own messages to achieve a deeper understanding of taken-for-granted norms in their societies. The decoding process is a fundamental element of critical pedagogy. Decoding, Freire (1970, 105) writes, "requires moving from the abstract to the concrete." To do this the educators developed a method we called *video reflection*, which will be discussed below.

The educators and I established the praxis in two very different locations in Nepal: Godamchaur and Kapan. Godamchaur is an agricultural village comprised mostly of *Dalits*, or untouchables, the lowest stratum of the Hindu caste system. Out of eleven participants only two were men in Godamchaur. Kapan, meanwhile, is a growing neighborhood in northwest Kathmandu with a great deal of recent urban development and increasing migration from rural areas. The participants in Kapan were largely from working class families. Four men joined the Kapan course along with six women. These were areas in which the educators were known by the local people and could gain access as insiders. As mentioned, my role was limited to observations of the classes in both settings and regular reflection meetings with the educators, during which I would give them feedback on their progress and help them prepare for their next class.

The *Insider Windows* praxis was an iterative process I lay out below, with the steps of problem posing, coding, participatory investigation, and decoding and reflection. Each of the two groups of participants went through three iterations of the steps, totaling twelve sessions in each setting before the course reached its conclusion.

The first three sessions of the class were dedicated to camera and movie production training. I provided each of the educators with a video production kit to use in their respective setting.[2] Participants played practice games borrowed from the participatory video[3] literature (Lunch and Lunch 2006) in addition to games developed for the course. The purpose of these games was entirely pedagogical. They involved fun "learning by doing" activities that gave the novice participants the opportunity to interact with the cameras and learn the various technical functions.

The transition to more critical and reflective discussions occurred at the fourth meeting, which contrasted with the lighthearted atmosphere of the first three sessions. This portion of the class began with a topical theme introduced by the educator. The topical theme was drawn from the commonalities found in newspaper articles that the participants were asked to bring in and share with the group. In this way it was indirectly elicited from the participants.

In Godamchaur, all the participants except one neglected to bring in a newspaper article so the educator asked them to discuss *responsibility*, and she selected that as the topical theme. She asked the participants what responsibilities they had fulfilled that day and what effects, if any, does shirking one's responsibilities have on their community or society. Participants responded that a family, for example, is affected if lunch is not prepared, because family members will go hungry. Another participant said that if the Village Development Committee issues a budget allocation to a contractor to fix a road and the contractor does not do it, then the whole village would suffer.

Meanwhile, in Kapan, the educator identified child-related issues as a prevailing theme in the articles her participants chose. She introduced the theme *children* to her group. In this stage of the process, an educator poses questions that investigate or problematize the themes being discussed by the class (Freire 1970; Shor 1992). To this point, Morrow and Torres (2002, 121) write, "The general pedagogical principle involved here is *active participation* and *not* that teachers should never have expertise to communicate or that classroom interaction should always and everywhere take the immediate form of a dialogue [italics in original]." Through the process of problem posing guided by the educator, participants critically investigated and interrogated the themes based on their own contexts and in the language of their daily vernacular.

Once a theme had been problematized and discussed, participants coded it by storyboarding, which is also an important step in the participatory video process (Walker and Arrighi 2013). Storyboarding involved dividing the class into subgroups, which consulted together to create a narrative that best represented the theme they had chosen collectively as a class. Participants would draw the narrative in picture form, almost like a comic book. As a result, storytelling became the first point of knowledge creation in the praxis. In Godamchaur, for example, the participants generated the theme *overcoming negative forces*, and a subgroup of the class represented that theme by creating a story about coming together as a community to collect litter in the village. These storyboarded themes were then recorded with the video camera, often using actors and scripted dialogue, which was the creative investigation of the theme. The movies were then screened back to the entire class to be analyzed and decoded by all the participants in a reflection process that came to be known as *video reflection* (See Figure 13.1 for a graphical representation of the praxis).

During the *video reflection*, participants broke up into their production subgroups and devised questions to pose to the filmmakers from the other

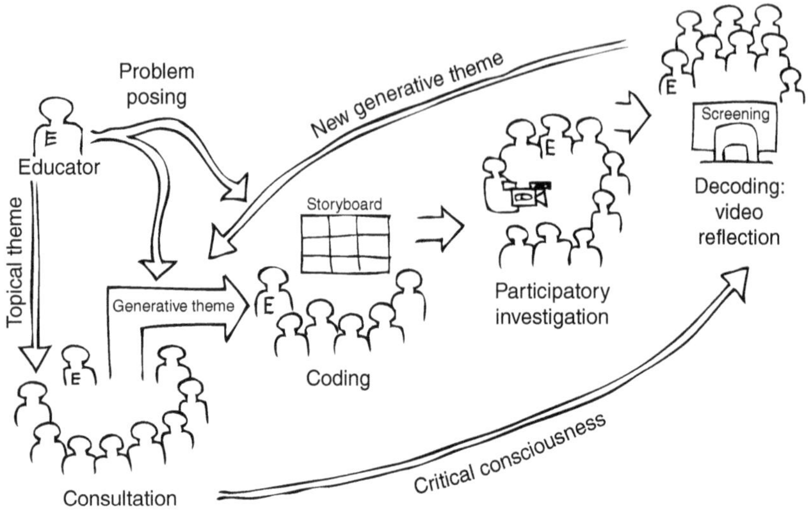

Figure 13.1 The critical pedagogy paradigm using movie-making as praxis.
Source: author.

subgroups. Questions often centered on the reasons why a subgroup chose to represent a theme in a particular way, and what certain scenes were intended to depict. These discussions were recorded in a TV show style format, meaning one subgroup interviewed another, with members acting in different roles such as anchor, producer, and cameraperson. The recorded segments were then screened for the class, and the participants had the opportunity to see themselves and their peers answering the questions. This opened the door to self-reflexivity, which, as identified by Crocker (2003), is an element more complex than the relatively straightforward concept of empowerment through communication. Self-reflexivity, through the experience of seeing oneself and one's reality on screen, promotes "feelings of confidence, self-worth, [and] better self-image, all as a result of seeing yourself as others see you" (ibid., 130). About her group's movie, one of the participants stated:

> We showed the problem from the perspective of the common people; we did this by understanding the problems faced by the common people. In truth, we ourselves are common people, so we understood that these problems are ours, and we made this movie based on problems we ourselves face.

While coding a theme was initially descriptive, decoding became very analytical. The educators in this study quickly recognized that abstract themes such as *overcoming negative forces* were partly solidified through narrative creation—a component of the coding process. They also recognized that decoding would have to be an analytical process, a counterpoint to the often descriptive nature of coding. Watching and discussing the representation of abstract themes, and seeing none other than their peers—indeed, themselves—acting in the stories, made the total transfer from abstract to concrete possible.

Out of this self-reflexive and analytical process, new generative themes emerged for discussion. The process began again and the next themes to be problematized and storyboarded were more sophisticated, but still generated dialogically by participants. For example, in Godamchaur, out of the discussion of the movies representing the theme *responsibility*, the theme *overcoming negative forces* was generated. The value of the films was their instrumentality in propelling the praxis forward. This is what differentiates this form of participatory media from direct media that have a goal of larger audiences and widespread dissemination. There was no final outsider destination for the films outside of the class meetings unless the participants negotiated one as part of a planned intervention that emerged from the praxis.[4] Such an open-ended model based on process is, according to Morrow and Torres (2002, 125), "oriented toward learning how to *produce* knowledge." The production of knowledge through this dialogic and investigative process is conscientization, and is, in fact, the goal of the exercise. This new critical consciousness allowed the *Insider Windows* participants to see the relationship between their own experience and social issues, "between single problems and the larger social system" (Shor 1992, 127).

Because this praxis is based on the use of generative themes, it is important to remember that participants can focus their inquiry on any point of their choosing on the knowledge horizon. The role of the critical educator, and his or her problem posing, is simply to ensure that the praxis is directed toward a more socially and environmentally just world. If that sounds ambiguous there is a reason—it ensures that the integrity of true subjectivity on the behalf of knowledge producers remains intact. While the educator can propose topical environmental themes at the beginning of the praxis, that act alone is no guarantee subsequent generative themes will remain on that trajectory. For example, an environmental topical theme such as *deforestation* could conceivably result in participants—after a coding, investigation, and decoding process—generating a non-environmental higher-level theme about the *limitations of the central government*. The counterpoint to this is that when participants generate environmental themes out of non-environmental topical themes, they are an expression of local, environmental value constructions.

Movies on environmental themes

The subject matter of movies produced by *Insider Windows* participants over the course of their twelve-week praxis was remarkably diverse. Issues addressed included teenage suicide, discrimination, and labor migration. The single most common thread, however, was the environment. While several films were entirely about environmental issues faced by participants in their communities, such as pollution, waste disposal, and water rights, environmental messages were also inserted into unrelated narratives in the form of asides. For example, in a film about inter-caste marriage, participants scripted the dialogue in such a way that actors would comment on the foul smell of garbage polluting the streets of their village, or scold one another when they littered. The following synopses are of three films produced by participants that dealt entirely with environmental issues.

Dharo Bhitrakaa Kathaaharu (Stories from the Water Tap)

The *dharo* (water tap) is the public water resource, and in this movie participants re-enacted a common story of conflict that occurs regularly in the village. The movie begins with a group of women going down to the water tap and bemoaning the recent water shortages. One complains that she does not have enough to water her crops. They await the arrival of the water flow for an hour, and when the tap finally begins to function there is a disorganized rush while women attempt to fill their vessels. The scene descends into chaos and arguing. One woman tries to connect a hose to the tap to water her crops while the other women wait to fill their buckets for their households. While she is away they disconnect her hose in order to use the tap. She returns, furious, and the quarreling begins to escalate into a pushing match. Two young school-aged girls arrive and try to reason with all of the women amidst the chaos. Eventually, their voices of reason prevail, and all the women agree they should share the community water resource cooperatively by waiting in line, taking turns, prioritizing household use, and not taking more than is reasonable for one's needs.

Education Can Change the World

This film contains both documentary and narrative elements. The subtitle for the film is "There is no time to lose," and its theme is pollution in the village. Some villagers have fallen ill from disease spread by garbage. A local man initiates a grassroots awareness campaign that results in better habits on the part of the villagers. After the campaign, there is a scene in which a young girl scolds one of her friends for littering. The narrator explains that the mindset of the people in the village has changed. In the final scene one of the participants is shown sweeping the small village temple before the narrator appears on camera herself, appealing to the viewers that cleaning the community begins with them, and there is no time to lose. They cannot depend on outsiders to manage the environment of their community for them.

Clean City is Healthy City

The third film is about the garbage problem in the Kapan area. A man who has been carelessly throwing garbage in a local informal dump becomes very ill. His son learns of a news article posted online about several people who have died in the nearby Balkhu area due to poison caused by surface garbage. He tells his father about the incident and suggests that he might be ill for the same reason. His father regrets throwing garbage and tells his son about how he was rude to a stranger who advised him not to drop his garbage at the informal dump. He promises his son that he will stop polluting the neighborhood. Later he even stops others from doing the same, encouraging them to wait and hand their garbage over to the municipality truck that comes to collect it. A fellow he has asked not to litter is then shown asking another person not to litter, and this highlights the way good habits can spread in a community. The film ends with the man, now healthy, walking around his neighborhood and admiring how clean it has become. He says, "If everybody is this aware, then our country will reach new heights and be developed."[5]

The messages contained in these films possess a certain veracity that precludes their dismissal as merely representations of the pragmatic function of environmental communication. The messages were crafted by insiders in a community and directed toward other insiders as part of an analytical investigation. This made them immune to cooption or dilution by agenda-driven outsiders. I make this point as a general principle. Doubtless, there were no outsiders with an anti-environment agenda making inroads into either class, so the subjective power of their knowledge was never tested in this context. If one contrasts this method of environmental value creation with the subject-to-object transfer of knowledge found in mainstream society, the purely constitutive function of a generative theme-based praxis becomes apparent.

An emerging critical consciousness

By the conclusion of the praxis, the confidence and analytical ability of the participants had expanded remarkably. These are both key elements in an emergence of a critical consciousness. Discussing the course in an interview, one of the educators stated:

> When I talked to the participants after the class was over, they said that now they are very clear about many things. Before they used to ignore many things, but now they are very much conscious. They want to make people aware about these things. They want to share the movies they have made and encourage people to think more. They say that they have started thinking. It will make a change, even if a group of two or three people [begin thinking]. Maybe not all nine of them, but even if two or three of them think critically it will make a change, if not for all the people, then for her family, or for herself only.

When viewed through the Johari Window, the educator's conversations with participants after the course's conclusion reflect the desired trajectory of knowledge generated through conscientization. The transition from "ignoring many things" to being "very much conscious" is represented at the framework level by revealing some of what is blocked behind Window 4, and, in the process, enlarging Window 3. The desire to make other people "aware about these things" is represented by the enlargement of Window 1, open knowledge, by drawing from Window 3. So what was the subjective experience of some of the participants? The following are excerpts from participant interviews and written feedback in the form of reflections:

> I learned to speak confidently. [The course] helped me to be able to discuss with my friends better, to have better conversations. We have [also] seen changes in our behavior. Now when I buy candy I don't throw the wrapper on the street like I used to. I put it in my pocket. So these changes have influenced our behavior.
>
> (Manisha)

> We were busy in our life, and we didn't have time to notice the problems of our society. But when we joined this class we started to search the problems of society. We found lots of problems we hadn't noticed before, like an uneducated society, child labor, a dirty city, etc. We were involved in the research. Through videos we tried to show our societal problems in detail. We were part of these videos.
>
> (Poonam)

> During the shooting of *Clean City is Healthy City* my project team were in search of a bunch of waste for the first scene of the movie. I suddenly noticed waste near my home that I had never noticed. At that very moment something knocked my mind that said, "Why shouldn't I contribute to my society?" Then me and my friends formed a group to clean the neighborhood. Working in a team and seeing through the camera has boosted up my mind and the way I see things around. I don't want to die ten years early due to pollution and I also want to make people around me live green.
>
> (Sujan)

All of these statements reflect a subjective inward or outward change on the part of the participant. It is important to keep them in context. While a sampling of reactions from students enrolled in formal education could conceivably yield similar responses, the significant difference is these inward and outward changes did not originate as the result of studying a predetermined curriculum. It is vital to remember that the critical educator did not present any knowledge to the course apart from the skills related to the camera. Participants dialogically generated all thematic elements that comprised the learning content, and that is what makes their statements representative of conscientization, rather than simple awareness raising.[6]

Here let us again consider the two aforementioned functions of environmental communication: to be pragmatic and to be constitutive. Just as one function undergirds the other, the production of movies mirrors this codependence.

The movies themselves, *as artifacts*, are examples of subjective pragmatic ecocultural communication: they delivered student-generated messages to community members who were enlisted in their production, and to the remaining participants who screened them and then composed reflection questions. Meanwhile, *the process* of their production was operationally constitutive; this can be seen clearly in Sujan's quote above. Only through the process of producing a film about garbage did he notice the waste near his home, and it was at that moment that "something knocked [his] mind." The process engendered in him a desire to "help society." Through his inquiry, Sujan revealed a previously unnoticed truth about his context, and his desire to do something about it was the result of unblocking a subjective understating of that context. In that way, through the coproduction of knowledge, the praxis was *value-formative*. In other words, the process was functionally constitutive environmental communication, and it provoked in Sujan an emergence from a state of submersion to that of a change agent in thought and action.

Readers will undoubtedly note that *Insider Windows* followed, by design, a Freirian praxis of critical pedagogy redeveloped to accommodate the use of video narratives. When recreating this praxis, environmental educators are encouraged to remember Freire's emphasis on the freedom of reinvention. There may exist new and innovative ways in which to introduce topical environmental themes into a praxis. Perhaps narrative-based investigations of society can initially be replaced by theatrical representations of local natural resource use. Indeed, that is thoroughly topical, but further research is needed to test the extent that subsequent generative themes remain within the sphere of environmental communication or depart to other dominant social issues a community may face.

This case does illustrate, however, that, through the use of generative themes, salient environmental issues still come to the fore in contexts where they have an impact on day-to-day lives. Few, if any, contexts exist today where environmental issues are not among the most pressing, and it is therefore unlikely that a purely generative theme-based praxis will not uncover them.

Zooming out

So what does a critical pedagogy for environmental communication offer the broader environmental discourse? This answer is not far from what critical pedagogy in general has to offer the oppressed: a means by which participants can visualize a subjective intervention into reality. Conscientization is followed by an emergence from a reality in which men and women were previously submerged. Consider the masses of people who live their lives without any heed for the way their actions impact our environment, despite the abundance of pragmatic environmental communication messages they regularly encounter. These people are submerged in a reality from which they must emerge in order to transform. This transformation in historical reality is what Freire (1970, 109) describes as an intervention, the step that follows conscientization. The anecdotes from the previous sections, of Sujan and his neighborhood-cleaning group and the young women in Godamchaur who screened a film for sixty villagers, are examples of self-driven interventions into historical reality.

In its endeavor for systemic change, critical pedagogy holds the position that change cannot come from within the dominant system itself—a system fully capable of coopting mainstream environmentalism if precautions are not taken. Take the example of the concept of ecosystem services: it has been used as a tool to communicate the economic value of tasks performed by the environment in order to use the rationale of the marketplace to bolster environmental arguments. There is nothing wrong with this per se, but as long as the discourse remains confined within the limits of the current system, a new a priori mode of thought will remain out of sight beyond the horizon. This is bona fide hegemony: not controlling what is said but rather normatively establishing the framework, logic, and language within which communication must occur for it to be validated. Some mainstream environmentalism does acknowledge the pressing need for systemic change at the level of thought; it is what Speth (2008) calls *a new consciousness*. Accordingly he writes, "today's problems cannot be solved with today's mind" (ibid., 204).

Worth considering is the fact that many indigenous societies have very different historical a priori notions of nature, and these notions are in danger of being coopted by uncritical Western environmentalist discourse as long as it remains within the current framework. As Milstein notes (2011, 4), Western humans have an "insatiable desire to point and name" phenomena that they observe in the natural world. For many insiders, the natural world and everything in it has already been named and coded by outsiders, and attempts to attribute subjective meanings that contradict this hegemony are met with resistance.

For that reason, it is unfortunate that the de facto method of transferring environmental knowledge and ethics today remains from North to South, from core to periphery, often employing the machinery of international development and global foundations. Acknowledging that the messages themselves may speak to a post-colonial and anti-imperialist worldview, they remain messages disseminated by institutions occupying the same structural landscape as those that created the environmental crisis in the first place. Therefore, the peril lies not in their content, but in their perceived ownership. In other words, for the oppressed, for whom Freire developed his pedagogy, environmental values must be constituted locally. This often involves returning to what was once known in the past, or generating new knowledge and understandings—both of which remain hidden behind Window 4 and can be revealed through a praxis that facilitates the self-awareness-raising of self-knowledge, or conscientization. Indeed, whereas community-level generation of knowledge can be transformational, subject to object transfer of knowledge will always be hegemonic, regardless of whether the message is capitalist or environmentalist.

Returning the discussion to the university—there is a danger that environmentalism in general can become institutionalized within higher learning. It is crucial that a global environmental ethic, the development of which will progress over the years to come, does not emerge exclusively from the confines of the ivory tower or the meeting rooms of intergovernmental organizations. Pedagogy that has a track record of revealing voices and perspectives in the borderlands, where environmental degradation is most keenly felt, can be successfully integrated into the higher education classroom even if piecemeal initially. In order to effectively integrate blocked subjectivities into broad environmental discourse, via the window of open knowledge, it is the constitutive function of environmental communication that must be engaged.

Notes

1 Shor separates topical and generative themes from *academic themes* in pedagogy. Academic themes are grounded in the formal knowledge of experts or outsiders, and by design they did not factor into the praxis.

2 Each video production kit contained one Sony camcorder with built-in projector, one camera bag, one handheld microphone, one extra battery, one tripod, and a ledger of large sheets of paper with markers for storyboarding. The cost of the entire kit was under $500.

3 The concept of participatory video broadly captures the range of activities involving novices creating a film together collaboratively as an act of political or social expression or investigation. It is commonly used as a facilitated activity in international development.

4 In Godamchaur, three participants (all under 20 years of age) borrowed the camera after their final class to make another movie to screen for the village. They shot a sixteen-minute film about caste discrimination that engaged over thirty villagers as actors and extras. They projected it at the local school and over sixty villagers attended the screening, demanding that the film be played three times in succession. Following the screening, the filmmakers facilitated a discussion about issues raised in the film.

5 All three of these films can be seen on the project *Vimeo* channel: vimeo.com/channels/insiderwindows.

6 Discussions about the fundamental difference between these two concepts can be found in the literature, and one in particular is worth quoting at length. On the subject, Rahman (1991, 17) writes:

> The term "conscientization," which has been popularized by Paulo Freire, is widely used to refer to raising people's awareness. As Freire has made quite clear, conscientization is a process of self-awareness-raising through collective self-inquiry and reflection. This permits exchange of information and knowledge but is opposed to any form of teaching or indoctrination. But the practice of 'conscientization' often departs from this concept, and slips consciously or unconsciously into processes of knowledge transfer rather than the stimulation of an assistance to processes of the people's own inquiries to build their self-knowledge. This has nothing to do with conscientization, and in fact inhibits the development of self-awareness as well as the self-confidence needed to advance self-knowledge.

References

Au, W. (2007) "Epistemology of the oppressed: The dialectics of Paulo Freire's theory of knowledge," *Journal for Critical Education Policy Studies*, 5(2): 175–96.

Cox, R. (2013) *Environmental communication and the public sphere*, Sage Publications, Thousand Oaks.

Crocker, S. (2003) "The Fogo Process: Participatory communication in a globalizing world," in White, S., ed., *Participatory video: images that transform and empower*, Sage Publications, Thousand Oaks, pp. 122–41.

Darder, A., Baltodano, M. and Torres, R. D. (2009) *The critical pedagogy reader*, Routledge, New York.

Freire, P. (1970) *Pedagogy of the oppressed*, Continuum International, New York.

Freire, P. and Shor, I. (1987) "What is the 'dialogical method' of teaching," *Journal of Education*, 169(3): 11–32.

Giroux, H. A. (2009) "Critical theory and educational practice," in Darder, A., Baltodano, M. and Torres, R. D., eds., *The critical pedagogy reader*, Routledge, New York, pp. 27–51.

Giroux, H. A. (2011) *On critical pedagogy*, Continuum International, New York.

Illich, I. (1970) *Deschooling society*, Marion Boyars, London.

Kahn, R. (2005) "From Herbert Marcuse to the earth liberation front: Considerations for revolutionary ecopedagogy," *Green Theory and Praxis Journal*, 1(1): 71–99.

Lunch, N. and Lunch, C. (2006) *Insights into participatory video: A handbook for the field*, Insight, Oxford.

Macrine, S. L., McLaren, P. and Hill, D. (2010) *Revolutionizing pedagogy: Education for social justice within and beyond global neo-liberalism*, Palgrave Macmillan, New York.

Marcuse, H. (1972) *Counterrevolution and revolt*, Beacon Press, Boston.

Milstein, T. (2011) "Nature Identification: The power of pointing and naming," *Environmental Communication*, 5(1): 3–24.

Milstein, T. (2012) "Survive, critique, and create: Guiding radical pedagogy and critical public scholarship with the discursive guideposts of ecopedagogy," *Green Theory and Praxis Journal*, 6(1): 3–16.

Morrow, R. A. and Torres, C. A. (2002) *Reading Freire and Habermas*, Teachers College Press, New York.

Rahman, M. A. (1991) "The theoretical standpoint of PAR," in Fals-Borda, O. and Rahman, M. A., eds., *Action and knowledge: Breaking the monopoly with participatory action-research*, Apex Press, New York, pp. 135–45.

Shor, I. (1992) *Empowering education*, University of Chicago Press, Chicago.

Speth, J. G. (2008) *The bridge at the edge of the world: Capitalism, the environment, and crossing from crisis to sustainability*, Yale University Press, New Haven.

Walker, G. and Arrighi, J. (2013) "Participatory video as a catalyst for informal learning and expression: A review of a PV training in Uganda, 2012," *LEARNing Landscapes*, 6(2): 409–23.

14 Repair Cafés

Reflecting on materiality and consumption in environmental communication

Sigrid Kannengießer

Repair Cafés are new and widespread events in which people meet to repair their everyday objects together. While some people voluntarily offer help in the repairing process, others seek help. The idea is not to provide a "free service center" but to help people help themselves free of charge. In this article, Repair Cafés are discussed as a highly relevant and useful phenomenon for educational purposes in environmental communication which aims at dealing critically with issues of materiality and (over)consumption of media technologies, and their destructive environmental effects.

The Dutch foundation Stichting Repair Café claims to have invented the concept of Repair Cafés in 2009 (Stichting Repair Café n.d.). Whether it was the origin or not, Repair Cafés have spread all over Western European and North American countries within the last several years (see a map for many of the locations at http://repaircafe.org/en/visit/). While repairing is an old practice, Repair Cafés are new in that the repairing has become public and, therefore, the event itself provides an innovative setting for environmental activism as the activity can be used to educate, organize, and lobby for environmental protection. Participants want to repair their things to avoid buying new ones. People bring "new" media devices such as laptops, computers, and smartphones, but also very old radios or slide projectors. Participants are aware of the polluting and harmful production and disposal processes of media technologies and try to contribute to a sustainable society by repairing their devices.

Repair Cafés are a highly relevant phenomenon for research and teaching in environmental communication. Although we use, analyze, and teach about media technologies, we tend to forget these technologies are produced under environmentally destructive conditions with resources extracted under harmful circumstances. When buying new media technologies regularly and teaching about "hip" new media technologies, we also tend to forget what happens with our disposed technologies—which are mostly transferred to African countries where the e-waste does serious harm (see e.g., Kaitatzi-Whitlock 2015 on e-waste). Repair Cafés are projects that acknowledge these problematic production and disposal processes, criticize the increasing consumption of media technologies, and try to offer an alternative.

In this chapter, I discuss Repair Cafés as a phenomenon with which materiality and (over)consumption can be discussed critically in environmental communication classes. I show from my teaching experience that Repair Cafés can prove useful both in classes on environmental communication theory and in research-service-learning-based or practical seminars in which students can organize these events themselves.[1] The chapter is structured as follows: I first sketch the research field of repair and Repair Cafés and show the relevance for the field of environmental communication. Illustrating the perspectives of media and communication studies, I show how Repair Cafés can be used as a phenomenon to discuss materiality in environmental communication seminars, highlighting three relevant aspects of the material dimension of media technologies—production, consumption, and disposal—that become visible through Repair Cafés. I then present experiences of a practical seminar I taught in 2014 at the University of Bremen, Germany, in which participating students organized a Repair Café themselves. My teaching experience shows that, after discussing the environmental and social effects of media production, consumption, and disposal, seminar participants became very motivated to contribute to alternatives to the consumption of media technologies. Finally, I link the theoretical discourse and practical experiences to underline the relevance of Repair Cafés in environmental communication pedagogy.

Research on repair and Repair Cafés and relevance to environmental communication

Repair and Repair Cafés are mainly analyzed in technology and design studies: here, repairing is defined as "the process of sustaining, managing, and repurposing technology in order to cope with attrition and regressive change" (Rosner and Turner 2015, 59). It is analyzed as a political practice (ibid., 55–6) where Repair Cafés are perceived as "theaters of alternative industry" whose participants strive for social change in questions of egalitarianism and collectivity (ibid., 67). Moreover, repairing is also perceived as art (Jackson and Kang 2014) and analyzed in developing countries (Jackson et al. 2012). Also the gender roles, which are (de)constructed in these events, are analyzed (Rosner 2013).

For media and communication studies, and especially the field of environmental communication, repairing and Repair Cafés are highly relevant, since the repairing of media technologies can be understood as an alternative media practice designed to contribute to sustainability. The aims of many people bringing their media technologies are consumption-critical, meaning that the participants want to avoid buying new things by prolonging the life-span of existing devices (Kannengießer 2016, forthcoming-a and -b).[2] Organizers and participants explain they want to protect resources and avoid producing e-waste, as they are aware of the environmental pollution and the harmful social circumstances under which media technologies are produced and disposed (ibid.). The repairing of media technologies can be characterized as an example of consumption-critical media practices, that are "those practices which either use media to criticize (certain) consumption or are (conscious) alternatives to the consumption of media technologies such as the repairing, exchanging or production of durable media technologies" (Kannengießer 2016, 194).

Repair Cafés are a highly relevant phenomenon for the field of environmental communication, in which research and theory "focus on communication and human relations with the environment" (Milstein 2009, 344). Milstein distinguishes three interrelated aspects of research in environmental communication: material–symbolic discourse, the mediation of human–nature relations, and applied and activist theory (ibid., 346). The latter is "aiding somehow in social-environmental change" (ibid., 347). Since Repair Cafés are projects of environmental action, they are seen as part of the latter aspect within environmental communication, brought to life in pedagogical practice.

"As with other forms of communication, environmental communication is both an activity/phenomenon and a field of study that, not surprisingly, studies the activity/phenomenon" (Meisner, n.d.). The phenomenon of Repair Cafés can serve as both an activity that can be organized in practical environmental communication classes and a phenomenon that can be studied within theoretical classes in this field. As a practical project, Repair Cafés are examples of how environmental communication can be put into practice. This application is one ambition of this research field, as the journal *Applied Environmental Education & Communication* shows. Striving to apply environmental communication, publications in this field discuss principles and skills for practitioners (e.g., Jurin et al. 2010). Case studies show that environmental communication in practice is often perceived as the marketing or communication strategies of practitioners, such as environmental activists (see e.g., Parker 1997; McKenzie-Mohr et al. 2012). But the case of Repair Cafés shows that media practices, or the way media technologies are used and appropriated, can themselves be applied to environmental communication.

Repair Cafés can be used to discuss two key concepts of this research field: materiality and consumption. Research shows that, by repairing media technologies, people participating in Repair Cafés strive for sustainability, the protection of resources, and the avoidance of waste (see Kannengießer forthcoming-a and -b). Participants, therefore, acknowledge the materiality of media technologies, as they are aware of the problematic effects regarding production and disposal of these goods. Actors involved in these events know about the dangerous extraction processes of coltan in Democratic Republic of the Congo, a resource used in all digital media technology, and the harmful and illegal disposal of e-waste in the African continent (e.g., Ghana). By repairing media technologies, participants try to diminish production, as well as the disposal of existing devices.

In media and communication studies, materiality is discussed in different ways. Lievrouw (2014, 33–42) sketches the research field on materiality in media and communication studies and differentiates between four research strands:

- media effects studies have perceived media technologies as channels, which enable linear communication processes;
- in communication network studies, not only the technologies but also social relations such as actions, interactions, and contacts are viewed as material;
- medium theory shifts the focus from the content of the media to its form; and
- cultural media studies criticize these strands as techno-determinist and put the focus on the people who appropriate and shape technologies.

These research strands deal with the materiality of media technologies in the reception processes. Domestication theory does so, too, focusing on media appropriation processes within the home and noting that media technologies become "invisible" in these processes (e.g., Silverstone and Hirsch 1992).

Repair Cafés remind us about the before and after of the use or appropriation of media technologies. Richard Maxwell and Toby Miller (2012, 9–10) criticize media and communication studies as largely ignoring "the physical environmental effects of media." In media and communication studies we have to ask and be aware of where the media technologies we use and analyze come from and where they go after—whether broken or still working. To distinguish themselves from the "old" media ecology research (which used environment as a metaphor to grasp the connection between culture and technology), Maxwell et al. (2015, xiii) stress, there is a "new" media ecology, which "is committed to the materialist approach." This new media ecology analyzes the effects of (digital) media technologies on the environment. This approach sees the materiality of media technologies as a process (Gabrys 2015, 6), meaning that the moments of production, consumption, and disposal are perceived as dimensions of materiality that have effects on the planet. These three moments become visible through Repair Cafés, which can serve as a pedagogical starting point to go deeper into discourses of production, consumption, and disposal.

Production, consumption, and disposal of media technologies becoming visible and discussible through Repair Cafés

The motivations of people participating in and organizing Repair Cafés make three stages within media technologies processes visible: production, consumption, and disposal. In my research on Repair Cafés, participants wanted to avoid buying new media devices, and thereby overconsumption of these goods, as they knew about the polluting and socially harmful production processes. For instance, the organizer of one Repair Café said, "I think especially the repairing of computers is important as they contain resources that are directly related to the deaths of people in other countries. And we should not throw these [technologies] away and buy a new IPhone."[3] Within environmental communication, there are some texts that focus on such ecologically harmful production processes and their consequences for the environment (Bleischwitz et al. 2012;

Maxwell and Miller 2012; Chan and Ho 2008). In teaching the production processes of media technologies, these or similar studies can be discussed.

I find it is also useful to use media texts, such as newspaper articles, radio features, or (short) films to discuss the production processes of media technologies. News coverage deals with the extraction of coltan in Democratic Republic of the Congo, for example, or the production of technologies in sweatshops in Southeast Asia. As newspaper articles are often short texts, one possibility is to let students read and discuss different articles in groups within the seminar (one per group) and ask them to represent "their" article as well as their opinion about it.

Moreover, several films deal with the harmful production of media technologies. One very impressive film is *Blood in the Mobile: Mining in the Congo* by Frank Piasecki Poulsen (2011), in which the filmmaker went to the mines in Congo where coltan is extracted in dangerous and polluting ways. Showing this documentary, or parts of it, makes the environmental and social consequences of media technologies production visible.

A second relevant stage highlighted by Repair Cafés is consumption. One young participant in a Repair Café said he wanted to repair the laptop he brought because he was not a "consumer-person." And an organizer of a Repair Café said, "We cannot talk about sustainability without a culture of repairing, and a fundamental extension of the life-span of technologies." He further stressed that a culture of repairing is needed to establish a sustainable society:

> We need to have a cultural change, through which it becomes cool again and socially acceptable to walk around with technologies which have signs of use and patina, where the display has scratches or fractures and one says: "This is my good old device, I stand by this, this is my trademark."

Media and communication studies often refer to consumption as the consumption of media content: e.g., the TV series people watch and how often. Although the consumption of media content is often discussed critically (Canclini 2001), there are more dimensions to media consumption. Within environmental communication, the consumption of media technologies also becomes relevant in material ways. In mediatized societies, in which the media environment is becoming more and more complex and media are gaining in importance in all societal areas (Krotz 2009), the number of media devices is increasing. In consumer societies in so-called "(post)industrialized" countries, people are buying more and more devices more often: each household has not only one TV but many; people own not only one computer, but two or more; schools, restaurants, trains, planes, etc., are equipped with media technologies, and these devices do not "live" long but are replaced regularly. Jon Raundalen (2015, 118) argues media coverage of new media technologies drives consumption as "journalists are providing the industry with a lot of free advertising."

The organizers and many participants of Repair Cafés not only criticize this accelerated consumption of media technologies, but, with the repairing of technologies, they also are pursuing a consumption-critical media practice. Consumption-critique and consumption-critical campaigns are often examined in media content analysis within the field of political communication (e.g., Gaßner 2014; Baringhorst et al. 2010; Micheletti and Stolle 2007; Greenberg and Knight 2004). But consumption-critical media practices are less acknowledged and the repairing of media devices can be discussed as such an example from a perspective of environmental communication by problematizing the environmental and social effects of consumption and by stressing how to avoid these by repairing these goods (Kannengießer 2016). People in Repair Cafés do not criticize the use of media technologies fundamentally—they use them themselves—but they stress that a more sustainable way of appropriation and consumption is necessary.

Finally, the disposal of media apparatuses is a third aspect, which is brought to attention by Repair Cafés. One participant of a Repair Café pointed to the effects of reducing e-waste: "We would have a better world if more people repair their things … because our planet would be less polluted." Several studies deal with the disposal of media technologies and its social and ecologically negative effects (Kaitatzi-Whitlock 2015; Bily 2009; Robinson 2009; Widmer et al. 2005). Like the topic of production, disposal also has gained news coverage describing the negative aspects of the disposal processes, and these texts can provide context in environmental communication classes. Again, there are films that document the disposal process and show the situation in garbage dumps in Ghana, for instance, where people (often children) try to extract some of the material from disposed media technologies and in doing so damage their health. And the films shed light on the effects of e-waste disposal. One example is *Toxic City: German Waste in Ghana* by Christian Bock (2011).

My teaching experience is that, when harmful ecological and social effects of the production and disposal of media technologies are discussed in environmental communication classes, students are sensitized to consequences of overconsumption. Students in my seminars became concerned—most of them had not been aware of the circumstances under which media technologies are produced, and they also did not know what happened to their technologies after they threw them away. Once sensitized, they at first felt rather powerless when faced with this situation. They explained they had no choice but to use their computers, smartphones, television sets, etc., and only politicians could do something about this situation by, for example, changing laws. It was important to explore alternatives with them. Aside from the theoretical reflection, students can engage and work on alternatives by creating their own Repair Café as I show below.

Organizing a Repair Café as an alternative to overconsumption

Besides reflecting on Repair Cafés theoretically and discussing the relevance of materiality (with its aspects of production, consumption, and disposal), this phenomenon can also be a subject in practical seminars and research-service-learning. In the following, I present experiences from a practical seminar in 2014, in which students of the undergraduate program in media and communication studies at the University of Bremen organized a Repair Café as part of their course.

The main task for the students was to organize two Repair Café events. Before starting to organize the events, the students had to read texts on aspects of production, consumption, and disposal to contextualize the practical seminar within media and communication studies, and more specifically environmental communication. This sensitized students to the relevance of repairing media technologies and the idea of Repair Cafés. The Repair Café organized in this seminar took place in Café Noon in downtown Bremen[4] in a building administered temporarily by a non-profit organization that organizes the rent of unoccupied buildings for the city. We chose this location for the Repair Café as we thought the idea of using buildings that were not new and that were temporarily unoccupied would perfectly fit with the idea of not throwing away goods that can still be used.

One of the key tasks for the students was to find "experts" who would offer their help voluntarily at the event in repairing objects brought in by participants. It quickly became apparent that many of the students themselves could actually serve in this expert function for various kinds of items. In addition to being able to repair a bicycle, for example, some students proved to be competent in helping with the maintenance of laptops, smartphones, and radios, as well as sewing fabrics. Nevertheless, the participants gathered support through acquaintances, friends, and other "experts." In addition, the students developed a logo, designed flyers and posters, set up an event page on Facebook (www.facebook.com/repairbremen), and conducted public relations that led to local newspapers *Weser Kurier* and *Delmenhorster Kurier*, as well as the Bremen Tourist Center and Radio Bremen, reporting on the initiative before and after the events. There was also a report on the local television program *buten und binnen* and the student online magazine *Krosse*, a project of the University of Bremen.

It was mainly due to students' successful public relations measures that both Repair Cafés were very well attended. A variety of people brought defective everyday objects and devices that were (mostly successfully) repaired with the free assistance of the students and other "experts." In addition to repairing, the events emphasized communication among organizers and participants: not only did participants share knowledge about repair options and experiences, they also critically discussed issues such as planned obsolescence or the production and disposal of the specific things. Here, the students were facilitators in these discussions as they had read about and reflected on these topics before in the seminar.

After the first Repair Café, we reflected on the event in a seminar session and improved our organization of the second event—students noticed for example that there have to be more "experts" repairing electronic devices, which were the dominant goods brought to the first event. This reflection also made it clear that the students where highly motivated by the success of the first event.[5] After the two events, students had a reflection paper assignment. The task for this was to:

- describe and reflect on the organization of the Repair Café;
- contextualize the phenomenon of the Repair Café within media and communication studies and especially environmental communication; and
- reflect on the societal and personal relevance of the Repair Cafés.

The assignment served to challenge the students to combine theoretical reflections on the material aspects of media technologies and the organization of the Repair Café as an alternative to consuming media technologies.

One student explained her personal motivation regarding the organization of the Repair Café:

> On the one hand, I think it is reasonable to get broken things started again—because of the lack of environmental and financial resources. On the other hand, [Repair Cafés] provide space in a society characterized by stress and time pressure to learn something together, to share skills, and to become conscious: What do I consume, and how do these things work? How can I repair them and with whom? And do I need these things actually?

The student points out the relevance of Repair Cafés in exchanging repair knowledge and provoking moments of reflection regarding the consumption of media technologies. The latter aspect was mentioned by another student in her report: "Through events like Repair Cafés, sustainability and environmental issues become topics of conversation—people start to think about these issues." She highlights the relevance of Repair Cafés in provoking discussion, since people meet and not only repair their things but also debate issues like environmental protection and sustainability.

This same student in her reflection paper assignment also highlighted the relevance of Repair Cafés for social change:

> In my opinion, events such as these offer each individual the opportunity to make a small contribution to sustainability by not just purchasing a new radio when the old one can still be got going again. Maybe a Repair Café cannot make a significant difference, but imagine that more and more Repair Cafés are established all over the world, then the phenomenon of Repair Cafés can be perceived as something big, which *can* make a difference.

Another student reflected on the democratization potential of Repair Cafés and pointed to the empowering effects these events might have:

> I think it is important that a dynamic of democratization develops in which people, regardless of their origin, class, or status, are empowered to deal with everyday technologies competently and confidently … Repair Cafés serve as institutions to give people space to deal with technologies independently and self-determinedly, to explore them, infiltrate them, and empower themselves regarding their devices. I think these developments are important so as to not leave technologies to specialists.

These examples of student reports show ways the organization of the Repair Cafés helped students reflect on the relevance of these events for society, and their own efficacy in contributing to knowledge exchange, sustainability, and democratization.

Relevance of Repair Cafés in teaching environmental communication

Repair Cafés are a phenomenon with which materiality and (over)consumption can be brought up in environmental communication classes and production and disposal of media technologies can be critically discussed. Moreover, Repair Cafés can serve as experiential examples of alternatives to consuming media technologies as students organize Repair Cafés as part of their environmental communication classes.

Participants in Repair Cafés are trying to avoid the wasteful cycle of production and disposal of their media devices by prolonging the life-span of their apparatuses. The repairing of media technologies can be understood as a consumption-critical practice with the overall aim of sustainability. This motivation reminds us within media and communication studies in general and environmental communication in particular to take the material aspects of media technologies we use and teach about into account. In discussing and organizing Repair Cafés in environmental communication classes, students are sensitized regarding dangerous and harmful production and disposal processes and their own (over)consumption of media devices.

Repair Cafés are not the only example through which consumption or materiality can be brought up in environmental communication classes. There are many other examples within the do-it-yourself (DIY) movement that criticize overconsumption and analyze the problem of materiality, including Transition Towns, Urban Gardening, and exchange/swap circles (where goods are exchanged and shared to avoid disposal and consumption). In all this, environmental activism, media, and communication become relevant in the use of social networking sites, blogs or online forums to connect and mobilize people in the use of public relations with news media, and in the ecocultural discourses that are disrupted, challenged, and reconfigured. These various projects and more can be analyzed and organized in environmental communication classes, making the materiality of their environmental effects visible and accessible.

Notes

1 I distinguish here between theoretical and practical seminars according to the curriculum of the bachelor classes in media and communication studies at the University of Bremen, where I taught the seminars focusing on the Repair Cafés I discuss in this chapter. Theoretical seminars are those courses in which mainly academic texts are discussed. Practical seminars are courses in which students conduct projects.

2 In a qualitative study in 2014 and 2015, I analyzed Repair Cafés in Germany using the research questions:

- Why do people participate in Repair Cafés?
- What do the Repair Cafés and the practice of repairing mean to the participants, especially when repairing media technologies?
- And what societal relevance do participants see in the Repair Cafés?

I used grounded theory to pursue answers to these questions and conducted observations as well as interviews with organizers, people offering help in these events, and participants seeking help. The results of the study, funded by the University of Bremen, are published in Kannengießer forthcoming-a and -b. It was through this research that I became motivated to organize a Repair Café myself, although I am not an expert in repairing. I thought that organizing it in a seminar context at the university would be a great opportunity to sensitize students regarding the aspects of materiality and consumption of media technologies and to help us all realize alternatives to overconsumption and experience that we can make a small difference.

3 I translated German quotes from my research project and from student assignments into English.

4 Bremen, a city of 500,000 inhabitants, is located in northwest Germany.

5 After these two events, the Repair Café continued, now taking place at *Café Karton* in Bremen (www.facebook.com/pages/Karton/977229215626347) and organized by a group of people including some of the students who took part in the seminar.

References

Baringhorst, S., Kneip, V., März, A. and Niesyto, J. (2010) *Unternehmenskritische Kampagnen*, VS, Wiesbaden.

Bily, C. A., ed. (2009) *What is the impact of e-waste?*, Greenhaven Press, Detroit.

Bleischwitz, R., Dittrich, M. and Pierdicca, C. (2012) "Coltan from Central Africa, international trade and implications for any certification," *Resources Policy*, 37: 19–29.

Bock, C. (2011) *Toxic City: German Waste In Ghana* [film]. Available at: https://www.youtube.com/watch?v=kTWwsDAl4jE (accessed July 31, 2015).

Canclini, N. G. (2001) *Consumers and Citizens: Globalization and multicultural conflicts*, trans. and with an introduction by George Yudice, University of Minnesota Press, Minneapolis and London.

Chan, J. and Ho, C. (2008) *The dark side of cyberspace: Inside the sweatshops of China's computer hardware production*, Berlin. Available at: http://goodelectronics.org/publications-en/Publication_2851/ (accessed January 13, 2014).

Gabrys, J. (2015) "Powering the digital: From energy ecologies to electronic environmentalism," in Maxwell, R., Raundalen, J. and Vestberg, N. L., eds., *Media and the ecological crisis*, Routledge, New York, pp. 3–19.

Gaßner, V. (2014) "GreenAction: Die Kampagnen-Community," in Voss, K., ed., *Internet & Partizipation. Bottom-up oder Top-down? Politische Beteiligungsmöglichkeiten im Internet*, Springer, Wiesbaden, pp. 129–48.

Greenberg, J. and Knight, G. (2004) "Framing sweatshops: Nike, global production and the American news media," *Communication and Critical/Cultural Studies*, 1(2): 151–75.

Jackson, S. and Kang, L. (2014) *Breakdown, obsolescence and reuse: HCI and the art of repair*. Available at: http://sjackson.infosci.cornell.edu/Jackson&Kang_Breakdown ObsolescenceReuse%28CHI2014%29.pdf (accessed September 4, 2016).

Jackson, S., Pompe, A. and Krieshok, G. (2011) "Things fall apart: maintenance, repair, and technology for education initiatives in rural Namibia," Proceedings of the 2011 iConference, Seattle, WA.

Jurin, R. R., Roush, D. and Danter, J. (2010) *Environmental Communication: Skills and principles for natural resource managers, scientists, and engineers* (2nd edn.), Springer, Dordrecht.

Kaitatzi-Whitlock, S. (2015) "E-waste, human-waste, inflation," in Maxwell, R., Raundalen, N. and Vestberg, L., eds., *Media and the ecological crisis*, Routledge, New York, pp. 69–84.

Kannengießer, S. (2016) "Conceptualizing consumption-critical media practices as political participation," in Kramp, L., Carpentier, N., Hepp, A., Kilborn, R., Kunelius, R., Nieminen, H., Olsson, T., Pruulmann-Vengerfeldt, P., Trivundža, I. T. and Tosoni, S., eds., *Politics, civil society and participation: Media and communications in a transforming environment*, Tartu University Press, Tartu, pp. 193–207.

Kannengießer, S. (forthcoming-a) "Communicative figurations in Repair Cafés," in Hepp, A., Breiter, A. and Hasebrink, U., eds., *Communicative figurations: Rethinking mediatized transformations*, Palgrave, London.

Kannengießer, S. (forthcoming-b) "Repairing of media technologies as unconventional political participation," in Wallner, C., Wimmer, J. and Schultz, K., eds., *(Mis)Understanding Political Participation*, Routledge, London.

Krotz, F. (2009) "Mediatization: A concept with which to grasp media and societal change," in Lundby, K., ed., *Mediatization: Concept, changes, consequences*, Peter Lang, New York, pp. 21–40.

Lievrouw, L. A. (2014) "Materiality and media in communication and technology studies: An unfinished project," in Gillespie, T., Boczkowski, P. J. and Foot, K. A., eds., *Inside technology. Media technologies: Essays on communication, materiality, and society*, MIT Press, Cambridge, MA, pp. 21–51.

McKenzie-Mohr, D., Lee, N., Schultz, P. W. and Kotler, P. (2012) *Social marketing to protect the environment: What works?*, Sage Publications, Los Angeles.

Maxwell, R. and Miller, T. (2012) *Greening the media*, Oxford University Press, Oxford.

Maxwell, R., Raundalen, J. and Vestberg, N. L. (2015) "Introduction: Media ecology recycled," in Maxwell, R., Raundalen, J. and Vestberg, N. L., eds., *Media and the ecological crisis*, Routledge, New York, pp. x–xxi.

Milstein, T. (2009) "Environmental communication," in Littlejohn, S. W. and Foss, K. A., eds., *Encyclopedia of communication theory*, Sage Publications, Los Angeles.

Meisner, M. (n.d.) "What is environmental communication?" Available at: https://theieca.org/what-environmental-communication (accessed July 5, 2015).

Micheletti, M. and Stolle, D. (2007) "Mobilizing consumers to take responsibility for global social Justice," *The annals of the American academy of political and social science*, 611: 157.

Parker, L. J. (1997) *Environmental communication: Messages, media and methods: A handbook for advocates and organizations*, Kendall/Hunt Publishing, Dubuque.

Piasecki Poulsen, F. (2011) *Blood in the Mobile: Mining in the Congo* [film], Koncern TV-og film production/Chili Film/Gebrueder Beetz film production.

Raundalen, J. (2015) "Tech support: How technological utopianism in the media is driving consumption," in Maxwell, R., Raundalen, J. and Vestberg, N. L., eds., *Media and the ecological crisis*, Routledge, New York, pp. 99–119.
Robinson, B. H. (2009) "E-waste: an assessment of global production and environmental impacts," *Science of the total environment*, 408: 183–91.
Rosner, D. K. (2013) "Making citizens, reassembling devices: On gender and the development of contemporary public sites of repair in Northern California," *Public Culture*, 26(1): 51–77.
Rosner, D. and Turner, F. (2015) "Theaters of alternative industry: Hobbyist repair collectives and the legacy of the 1960s American counterculture," in Meinel, C. and Leifer, L., eds., *Design thinking research*, Springer International Publishing, Heidelberg, pp. 59–69.
Silverstone, R. and Hirsch, E., eds., (1992) *Consuming technologies: Media and information in domestic spaces*, Routledge, New York.
Stichting Repair Café (n.d.) "About Repair Café." Available at: http://repaircafe.org/about-repair-cafe/ (accessed July 31, 2015).
Widmer, R., Oswald-Krapf, H., Sinha-Khetriwal, D., Schnellmann, M. and Böni, H. (2005) "Global perspectives on e-waste," *Science of the Total Environment*, 25: 436–58.

15 Cultivating pride

Environmental engagement and capacity building in the UTEP–Rare partnership

Carlos A. Tarin, Sarah D. Upton, Stacey K. Sowards, and Kenneth C. C. Yang[1]

Rare is an international environmental nongovernmental organization that works to develop conservation and sustainability-related initiatives in developing tropical and subtropical nations. The organization's signature outreach strategy, Pride campaigns, is a unique model of community engagement that seeks to cultivate relationships between local communities and Rare's implementing partners. These campaigns are tailored to create solutions to environmental problems in particular regions sensitive to environmental degradation. Rare launched its first regional training program in 2001 in Great Britain to train English-speaking conservation workers from tropical and subtropical locales. Regional centers have since been established in Mexico, Indonesia, China, Brazil, Mozambique, the Philippines, and Micronesia. Rare has trained over 214 local campaign managers in its Pride conservation campaigns to inspire local stakeholders to "take pride in their natural resources—that alter the way their communities relate to nature" ('#56—Rare' 2012). Rare has forged a strategic partnership with the University of Texas at El Paso (UTEP) that has resulted in the development of a formalized conservation communication curriculum through a specialized master's degree program that is used to develop and implement Pride strategies. In this essay, we focus on the UTEP–Rare partnership and Pride campaigns to explore how conservation pedagogy is employed to foster meaningful connections between campaign managers and local communities and how such pedagogy might be replicated in environmental communication programs or classrooms.

As focal points of Rare's outreach efforts, Pride campaigns allow for an interrogation of how praxis-oriented pedagogy can achieve tangible results while simultaneously informing organizational practices. Pride campaigns are developed, implemented, and assessed over the course of the two- to three-year program as members of the organization receive training on a variety of topics ranging from social marketing theory to qualitative and quantitative research methods to rhetoric. After partner organizations agree to partner with Rare, they nominate a member of their staff to participate in the program designed to implement a Pride campaign. Campaign managers are members of the partner organizations who receive training at one of Rare's regional headquarters. The curriculum for training Pride campaign managers is standardized globally, but allows for some flexibility in the development of outreach strategies that are intended to reflect the particular social and ecological context of each campaign site. Over the course of the implementation period, campaign managers receive no fewer than fifteen weeks of full-time university training that provides instruction on identifying barriers for participation, advanced project management, conflict resolution, marketing and media relations. Each phase of university instruction is punctuated by fieldwork phases that allow campaign managers to survey local community members, establish a comprehensive strategic plan for campaign implementation, and assess behavioral and attitudinal shifts (ibid.). In their study of Rare's Pride model, Hayden and Deng (2013) found that the campaigns were successful at significantly altering baseline knowledge and attitude objectives; the campaigns, in other words, have a clearly demonstrated success record for achieving conservation results.

Starting in 2008, Rare initiated a strategic partnership with the Department of Communication at UTEP that continues to serve multiple purposes within the organization. Most participants in the Pride training program are eligible to receive a master's degree from the university upon successful completion and implementation of a Pride campaign and coursework. The UTEP–Rare partnership serves two major pedagogical functions:

- The partnership incentivizes the Pride program that is used for the recruitment and retention of campaign managers. Because Pride campaigns require an extensive two-year commitment from implementing partners and campaign managers, the incentive of an advanced degree has been used to acquire talent for the organization. In this regard, Rare is unique from other conservation organizations because it provides a tangible incentive for members who join and commit to the campaign implementation process.
- UTEP also serves an important capacity-building source for the organization and its members by working to improve leadership skills and organizational ability to work with communities. Because faculty members are expected to visit the regional offices and routinely provide instruction, the UTEP–Rare partnership provides an invaluable source of expertise to the organization.

The UTEP–Rare program highlights the importance of scholarship focused on the increasingly central role of environmental non-governmental organizations (ENGOs) in fostering meaningful, long-term environmental engagement. As Cox (2007) notes, environmental communication ought to be driven by scholarship and theorization that works to connect academics, practitioners, and local communities in ways that meaningfully improve environmental conditions globally. Rare's approach to conservation is a clear departure from other global mid- or large-size ENGOs (Bryant 2002), which typically use a "one size fits all" approach to conservation messaging. By focusing on capacity building within local communities, the campaigns shift from advocacy to empowerment, and thus, facilitate a more sustained form of environmental engagement.

Additionally, numerous researchers have called for scholarship on the effects and importance of communication pedagogy in the context of real world practice (Canary et al. 2014; Kedrowicz and Taylor 2013; Murphy 2004). In focusing our analysis on the pedagogical assumptions that underscore Rare's Pride program, we aim to highlight the salience of a praxis-based curriculum that teaches "students how knowledge of and competence in communication are necessary for participation in the global community" (Kedrowicz and Taylor 2013, 99). Such a focus is vital for contextualizing the role that ENGOs will continue to play in the global environmental conservation movement.

This chapter uses the UTEP–Rare partnership as a case study for examining how the goal-oriented pedagogy inherent in the Pride program can be used as an operational framework for environmental practitioners and theorists. In the rest of this chapter, we outline central tenets of Rare's Theory of Change (ToC) model that is used to orient pedagogical and programmatic outcomes for campaign managers participating in this program. Given its uniqueness, we contend that the UTEP–Rare partnership offers two key pedagogical insights:

- the emphasis on environmental outcomes structures programmatic and curricular development in such a way that learning and training become fundamentally rooted in applied practice; and
- in offering a co-constitutive model for campaign development, the program enables the development of long-term capacity building that is inextricably linked to local communities through grassroots organizing.

The goal orientation and emphasis on capacity building are, we argue, important pedagogical approaches that can be extended beyond the UTEP–Rare program to a variety of organizational and educational contexts. We conclude this chapter by detailing the replicability of these key insights in other contexts.

Rare and the theory of change

Pride campaign managers in the UTEP–Rare program participate in university phases that merge theory, research, and practice to achieve barrier removal for measureable conservation results. Each phase, ranging from three to ten weeks, involves a combination of formal classroom instruction and hands-on programmatic development. In the first university phase students learn how to develop their campaign strategy and create a research plan, which they then take into the field to conduct qualitative and quantitative research, and update their campaign strategy. When students return for the second university phase with community-based knowledge, from the unique context where they work, they reflect on and revise their initial campaign strategy and create a targeted marketing strategy. After conducting their second phase of fieldwork, students return for the third and final university phase where they work to analyze and present the results of their campaigns. To effectively achieve desired conservation results, Rare offers a theory of change (ToC) (see Figure 15.1).

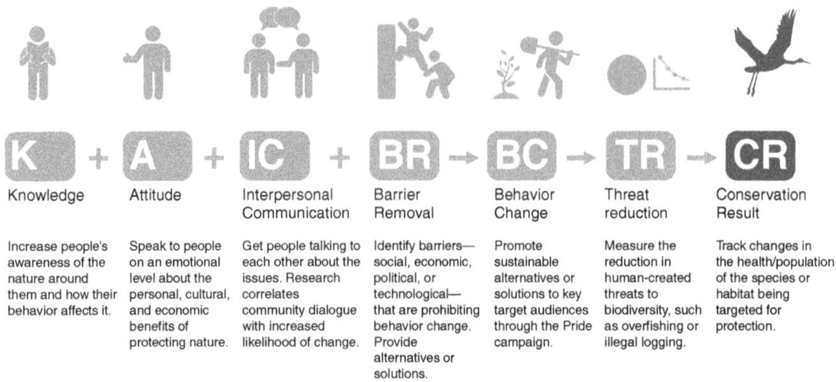

Figure 15.1 Rare's theory of change.

During university phases, campaign managers are given qualitative and quantitative research tools to help them assess knowledge, attitudes, interpersonal communication, and barriers already present in their local communities so that every part of the Pride campaign is guided by research-driven decision-making. As they prepare to begin the research process campaign, managers are given handouts and resources for determining research questions and approaches in addition to classroom lectures on research methods, conservation biology, social marketing, and leadership skills. Through in-class discussion and activities, they work with their instructors and each other to tailor research questions and methods to their unique target communities. For example, during the university phase campaign managers work to create a baseline survey for their fieldwork to measure community knowledge, attitudes, interpersonal communication, and behaviors.

Based on their initial assessment of community knowledge, attitudes, interpersonal communication, and barriers, campaign managers then work to create a barrier removal plan and targeted social marketing materials and activities. During their second university phase, campaign managers are provided with lectures, worksheets, and activities aimed at identifying barriers to behavior change at their field sites. For example, in the second university phase, instructors for the Latin American region in the Guadalajara training location used an activity where campaign managers identified the benefits and barriers to adopting the desired behavior change in their field sites. Working to embody the perspective of their target audiences, campaign managers wrote down possible barriers and benefits on pieces of paper and worked to collapse the items into larger categories as a collaborative form of qualitative data analysis. Thus, Pride campaign managers worked to create a social marketing campaign informed by their qualitative and quantitative data collection. Using information gathered to characterize their target audiences, campaign managers selected materials and activities for their Pride campaigns that were meaningful in their target communities, such as T-shirts, television and radio commercials, community events, social media pages, and much more, with the goal of raising awareness and creating behavior change.

The creation of the Pride campaign discussed led to behavior change, threat reduction, and a conservation result in the ToC as measured by qualitative and quantitative data collection before and after the campaign's implementation. Rare defines behavior change as the use of a Pride campaign to promote sustainable practice-based alternatives and/or solutions to key target audiences (Pride Campaigns n.d.). The conservation result is a recording of changes in the health and/or population of the animal species or natural habitat targeted for protection by the Pride campaign. For example, Pride campaigns may measure conservation results in terms of the growth of fish abundance and size in coastal fisheries or marine ecosystems, or the signing of reciprocal water agreements with community stakeholders to protect watersheds.

After their second round of fieldwork, campaign managers (i.e., students) complete their third and final university phase aimed at analyzing and presenting the results of their Pride campaigns. During this phase, campaign managers use quantitative and qualitative measures designed by Rare to assess how knowledge, attitudes, and practices have shifted as a result of the pride campaign, evidence of barrier removal and the adoption of new behaviors, and the long-term effects on threat reduction and conservation. This process is again completed in coursework through the use of lectures, activities, and discussions, and once their analysis is finalized, campaign managers present their findings and conservation results in a formal presentation, equivalent to the defense of a master's thesis.

The UTEP–Rare program ultimately uses active, engaged learning techniques during the university phases, extensive mentorship for the field phase, and the implementation of a conservation campaign based on extensive, community-based qualitative and quantitative research. Pride campaign managers learn about the ToC, qualitative and quantitative research methods, and social marketing, all informed by the unique conservation context in which they work. Because of UTEP–Rare's unique pedagogical approach, instructors, campaign managers, and local communities become active participants in how knowledge is constructed inside and outside of the classroom from formal education training to field work in data collection and community organizing. Campaign managers learn how to apply the knowledge gained in university courses to real-world conservation settings, and just as importantly, how to respond to local knowledge and conservation needs through the ToC approach. Importantly, the lessons emphasized by the Pride curriculum are intended to build personal, environmental, and organizational capacity. In other words, the emphasis on praxis is not limited to the campaign: Pride campaign managers frequently use their training when they resume normal duties with the implementing partners for whom they work. Finally, campaign managers learn how to identify, measure, and sustain the conservation result of their Pride campaigns. This conservation result is central to the program design and curriculum development and is tested through post-surveys that assess the impact of the campaign.

Outcome focus for curricular development

One of the key aspects of the UTEP–Rare program that makes it unique in comparison to many other communication programs is the focus on conservation outcomes. In designing the program curriculum, UTEP faculty and Rare staff discussed the best pedagogical and curricular tools needed to achieve specific conservation results. The curriculum was designed with three main outcomes in mind:

- to achieve a conservation result;
- to remove barriers within the communities that prevent such a conservation result; and
- to build capacity for conservation change both within the individual student (campaign manager) and the community, all of which result from the ToC model.

As part of the initial university phase, campaign managers must identify a key conservation threat within their communities. In Indonesia, for example, threats that campaigns have addressed include forest fires, illegal logging (and poverty connected to illegal logging), agricultural encroachment into protected areas such as national parks, overfishing, and coral reef destruction. In one of the most successful Rare campaigns in Indonesia, Eddy Santoso, the director of a local non-governmental organization (YAYORIN), in Central Kalimantan province, constructed his campaign around reducing forest fires in the Lamandau River Wildlife Reserve in order to protect forest habitat and a large population of orangutans. Forest fires in this area were largely the result of shifting agricultural practices in which fires were used to clear land. Using what he learned from university classroom training and Rare staff mentoring, Santoso worked with 197 out of 202 local farmers to establish more permanent agricultural practices, thereby reducing fire hotspots and forest fires significantly. The program trained Santoso and other campaign managers to identify a key threat, such as forest fires, and then to work with local community members to address that threat in a very specific and targeted way through the ToC model.

The pedagogical approach through ToC also requires that the campaign managers must identify the key barriers within their communities, which can be substantive. To overcome such barriers, campaign managers must find ways to engage community leaders and others within the community to not just change knowledge and attitudes, but also behaviors. One of the more successful global strategies learned through the UTEP–Rare curriculum has been to employ micro-credit unions that help local people to save a small amount of money that enables them more financial security, which in turn reduces their needs to engage in illegal logging or fishing in marine protected areas in the coral triangle region (Indonesia, Malaysia, and the Philippines). Several campaigns in Indonesia, for example, have focused on the use of such credit unions to reduce local community needs for fish. While the conservation result the campaign manager might be trying to achieve is focused on establishing marine protected areas or zones where fishing is not allowed, one of the ways to improve community buy-in is to determine why local community members would need to fish in protected areas. Many fishers work in protected areas because the more fish they can sell, the more income they can generate for their families. In rural areas where most of these campaigns operate, people are often close to subsistence fishers or farmers, so any income they can earn from selling fish is much needed, particularly in cases of emergency or unexpected expenses that arise. The micro-credit union enables community members to save extra income and borrow money when needed, meaning that fishers no longer need to fish to earn emergency income, thereby reducing barriers for effective conservation in establishing marine protected areas. The UTEP–Rare program focuses on how the campaign manager can effectively engage the community to set up such a micro-credit union as part of their campaign.

Finally, as part of the capacity building goal, the program and university curriculum focuses on building the leadership skills of the Pride campaign managers. Coursework includes extensive focus on how to identify key stakeholders, both formal and informal, and how to engage with such stakeholders, who might include political, customary, and religious leaders, as well as informal leaders who are respected within their communities. The curriculum trains campaign managers to build their confidence and include stakeholders in the planning process after the first university phase. Such planning creates buy-in and long-term relationships that enhance the effectiveness of the campaigns and barrier removals to achieve conservation results. Beyond extensive stakeholder meetings, campaign managers engage in social marketing techniques to increase awareness of conservation issues. Campaign managers learn how to use social marketing techniques, such as culturally appropriate religious messages, community events such as festivals and contests, cell phone text messaging, school visits, posters, banners, and other such messaging strategies to connect with those in their communities who can have an impact on conservation. For example, campaign managers in Indonesia have found that T-shirts are one of the most highly desired items and effective ways to disperse conservation information. In this sense, the pedagogical focus on campaign implementation is not just the invention of Rare and the individual Pride campaign manager, but also the work of the community as they co-constitutively create conservation goals that emerge into the campaign.

Co-constitutive campaign development

Rare's ToC and outcome focus are important pedagogical elements that Pride campaign managers learn and apply during their involvement with the organization. This conservation approach, however, is ultimately aimed at fostering meaningful connections between campaign managers, conservation organizations, and local communities through a process that centralizes the importance of co-constitutive engagement. The co-constitutive approach to communication education draws on the constructivist notion that "[c]ommunication is theorized as a process that produces and reproduces—and in that way constitutes—social order" (Craig 1999, 128). Because Rare Pride campaigns are planned for stakeholders with different cultural backgrounds, it is vital to examine the communication process in which meanings are negotiated and co-constituted between Rare, campaign managers, and local communities.

Pride campaigns aim to generate and establish community support to encourage sustainable behaviors ("A Rare Approach" n.d.) through a complex process of co-constitution among stakeholders residing in the campaign areas. Following Rare's ToC model, all Pride campaigns are developed using the social scientific Knowledge-Attitude-Practice (KAP) theory that is commonly used in social marketing campaigns. Rare's curriculum also emphasizes the SMART (Specific, Measurable, Attainable, Relevant, Time-bound) framework, which is designed to provide a step-by-step curriculum with measurable outcomes to educate environmentalists to conduct conservation initiatives around the globe. Accordingly, successful campaign outcomes can be accomplished through increasing stakeholders' knowledge and evoking emotional responses about the personal, cultural, and economic benefit of conservation ("Pride Campaigns" n.d.). According to this model, communication, dialogue, and conversation among people are used to cultivate subsequent behavioral changes through removing social, economic, technological, and political barriers to facilitate practical solutions (ibid.).

Rare's Pride campaigns can be conceptualized as co-constitutive insofar as they involve engagement with a wide variety of participants such as campaign managers, local stakeholders, program managers, volunteers, and government officials at each regional site. The outcomes of any Pride campaign depend on whether campaign managers are able to take into consideration many contradictory and dissonant communicative practices among stakeholders. Through the UTEP–Rare program pedagogical approach, campaign managers are provided with the analytical skills to analyze what constitutes threats to the environment, to collect both qualitative focus group data and large-scale behavioral data, and to apply evidence-based observations to assist campaign planning and outcome assessment.

The UTEP–Rare program aims to provide Pride campaign managers with intensive training that integrates stakeholders' decision-making processes into conservation campaigns that reflect social, cultural, political, and economic barriers in local communities. For example, stakeholders' concerns, attitudes, knowledge, and beliefs are identified through a series of focus group meetings, consensus building forums, and social scientific survey and interview data collection methods. The importance of resonating what stakeholders believe to be most critical and relevant to their own interests cannot be underestimated. Schoeneborn and Trittin (2013, 193) claim that corporate social responsibility programs are influential and effective "if they are resonant with other communicative practices ... and where practices of stakeholder involvement invite third parties to co-constitute these communicative boundaries." The emphasis on interacting with communities and stakeholders is a vital concern for all Pride campaign managers.

Previous Pride campaign successes demonstrate the salience of utilizing this co-constitutive model of message development to encourage behavior change to achieve conservation goals. For example, Baoyu Wei's campaign in the Dongting Lake Finless Porpoise Nature Reserve in China has integrated local cultural values into developing a campaign to increase the knowledge level of two groups of target audiences—fishers and villagers in the local communities—in order to cope with the diminishing number of fishery resources in Dongting Lake. The core slogan, "Conserving the Environment will Protect Our Descendants," effectively resonates with Chinese cultural values among local stakeholders and has proven to reduce overfishing behaviors, stop illegal fishing practices, raise awareness among local stakeholders of these problems, and equip local stakeholders to learn and take part in conservation efforts as a community.

In another Pride campaign in Perú, campaign manager Anselma Zumaeta Soplín focused on reforestation and conservation in the Rumiyacu and Mishquiyacu watersheds in Moyobamba, the capital city of the San Martín region. Through focus groups and surveys, Zumaeta Soplín identified housewives as major consumers of water in the region, and therefore a target audience of her Pride campaign. After her research showed that the most effective social marketing materials were advertising spots on radio and television, Zumaeta Soplín worked with community members to create advertisements starring local housewives giving advice about ways to conserve water while doing household tasks like laundry. Because they were co-created with local community members, these advertisements proved effective in both raising awareness and changing behaviors.

The success of this co-constitutive model of community engagement is not merely anecdotal; Pride campaigns consistently achieve tangible, measurable outcomes that reflect meaningful knowledge, attitudinal, and behavioral changes. For instance, the positive effects of family-oriented framing used in the Dongting Lake campaign are clearly observable in several effectiveness metrics before the campaign (pre-2011) and after the campaign (post-2012). Significant statistical differences between fishermen and villagers were observed between baseline and post-campaign effectiveness metrics. Most noteworthy, improvements were found in participants' knowledge level of being able to identify the Finless Porpoise as a keystone species that help to understand environmental threats in the Dongting Lake area.

By bringing local stakeholders into the campaign development and integration process, Pride campaign managers enact a form of co-constitutive environmental engagement that is reflexive and responsive to the needs and interests of local communities. Rare's emphasis on practical results, informed by the ToC, shapes organizational processes insofar as the Pride pedagogy is aimed at unearthing localized solutions to endemic environmental problems. Although the curriculum for the Pride program is relatively standardized, each campaign manager is afforded a relatively high level of flexibility in developing conservation messages that are tailored for their communities. Such an adaptive model of environmental engagement is necessary for addressing the particularities of different socio-cultural contexts and environmental exigencies that each campaign aims to resolve.

Conclusion: replicability and pedagogical extension

Rare's model of community engagement provides insightful case studies of the power and importance of adaptability within established pedagogical and instructional frameworks with a praxiological focus. Our chapter demonstrates the benefits of exploring pedagogical models of community engagement and organizing that seek to build capacity and empower local communities to create long-lasting change within their own communities. The strategies employed by Rare in engaging local communities can also be appropriated and translated to different pedagogical and strategic contexts.

While the focus on conservation results, barrier removal, and long-term capacity building are part of the UTEP–Rare programmatic curriculum that is unique to this partnership and each campaign manager's site, there are ways in which these insights could be replicated in a program or classroom. To adopt such an approach in the classroom over the course of a semester or quarter, using service-learning and final projects that focus on a conservation result, barrier removal, and capacity building on a smaller scale would be a great course-long assignment with a specific impact. For example, students could work individually or in teams to develop a semester-long campaign designed to address a conservation goal and barrier removal with community buy in. The campaign could focus on a campus-related problem, such as improving recycling, trash and waste management, water usage, driving and parking on campus, or energy usage in buildings. Students could then engage in identifying the conservation result they want to achieve, barriers that prevent such conservation, and a plan to develop stakeholder and community buy-in. This collaborative learning approach offers unique opportunities for students to understand the theories they are learning in class on a deeper level through engagement, and also teaches them to be responsive to the thoughts, interests, and actions of their communities. Similarly, classes might take an approach on some other conservation issue within the community or work with a non-profit organization that focuses on environmental or sustainability issues.

The UTEP–Rare program is a good example of how the salience and viability of exporting communication pedagogy and theory can work in specific conservation contexts. Initiatives such as Communication Across the Curriculum and Communication in the Disciplines have gained traction at many higher learning institutions in recent years, but these programs are still focused primarily at expanding the reach of communication skills in academic settings. We contend that engaged environmental communication pedagogy has an important, and even necessary, role to play in bridging the gap between scholars, practitioners, and local communities. The Pride program and the UTEP–Rare partnership are unique in their scope and emphasis, but the practical skills that campaign managers learn through the program can be integrated into virtually any conservation campaign or environmental communication classroom. Campaign managers, many of whom have no formal communication training prior to their involvement with the Pride program, routinely comment about how much even a basic understanding of environmental communication theory and practice has shaped their work outside of the Pride campaign. In this sense, the Pride program provides campaign managers with lifelong communication skills and resources that are used to promote individual, organizational, and community capacity. Models such as the Pride program underscore the potential of engaged environmental communication pedagogy to create meaningful community transformations.

Finally, Pride campaigns highlight the salience of using sometimes abstract environmental communication theory in practical domains. For example, disruption of the nature–culture dualism (Schutten and Rogers 2011; Sowards 2006) is not an explicit concern of Rare or the Pride campaign curriculum, but this concept can be clearly evinced throughout the organization and program. Rare's recently adopted mission statement, "To inspire change so people and nature thrive," highlights the importance of simultaneously considering the interests of ecosystems *and* human communities. Similarly, campaign managers are taught to understand the complex interconnections between human action and local ecosystems in ways that are meant to instigate a transformation of practices. Although many campaigns focus on curtailing activities that are environmentally harmful, many also include components that are meant to (re) stabilize local ecosystems. In this sense, Pride campaign managers "become spokespersons for other persons, birds, wolves, fish, trees, mountains, or any other community member in an attempt to promote community sustainability" (Peterson et al. 2007, 79). Rare's unique pedagogical approach and emphasis on praxis demonstrates the viability and importance of translating key environmental communication theoretical principles into action.

Note

1 Acknowledgements: Part of this research was funded by USAID grant number AID-497-A-12-00008-00, "Enhancing Behavior Change Through Conservation Programs."

References

"#56—Rare" (2012) *The Global Journal*, 23 January. Available at: http://theglobaljournal.net/article/view/556/ (accessed October 20, 2016).

"A Rare approach." Available at: www.rare.org/approach#.VIPCn9LF_vE (accessed May 2, 2016).

Bryant, R. L. (2002) "False prophets? Mutant NGOs and Philippine environmentalism," *Society & Natural Resources*, 15(7): 629–39.

Canary, H. E., Taylor, J. L., Herkert, J. R., Ellison, K., Wetmore, J. M. and Tarin, C. A. (2014) "Engaging students in integrated ethics education: A communication in the disciplines study of pedagogy and students' roles in society," *Communication Education*, 63(2): 83–104.

Cox, R. (2007) "Nature's "crisis disciplines": Does environmental communication have an ethical duty?," *Environmental Communication*, 1(1): 5–20.

Craig, R. T. (1999) "Communication theory as a field," *Communication Theory*, 9(2): 119–61.

Hayden, D. and Deng, F. (2013) "The science of goal setting: A practitioner's guide to goal setting in the social marketing of conservation," *Social Marketing Quarterly*, 19(1): 13–25.

Kedrowicz, A. A. and Taylor, J. L. (2013) "Engineering communication and the global workplace: Preparing professionals and global citizens," *Connexions: International Professional Communication Journal*, 1: 81–105.

Murphy, T. A. (2004) "Deliberative civic education and civil society: A consideration of ideals and actualities in democracy and communication education," *Communication Education*, 53(1): 74–91.

Peterson, M. N., Peterson, M. J. and Peters, T. R. (2007) "Environmental communication: Why this crisis discipline should facilitate environmental democracy," *Environmental Communication*, 1(1): 74–86.

"Pride campaigns." Available at: www.rare.org/pride#.VIPE69LF_vE (accessed May 2, 2016).

Schoeneborn, D. and Trittin, H. (2013) "Transcending transmission: Towards a constitutive perspective on CSR communication," *Corporate Communications: An International Journal*, 18(2): 193–211.

Schutten, J. K. and Rogers, R. A. (2011) "Magick as an alternative symbolic: Enacting transhuman dialogue," *Environmental Communication*, 5(3): 261–80.

Sowards, S. K. (2006) "Identification through orangutans: Destabilizing the nature/culture dualism," *Ethics and the Environment*, 11(2): 45–61.

Part IV

Environmental communication pedagogy and practice toolbox

This part provides a teaching exercise and activity toolbox for pedagogues based on teaching gifts of in-class and out-of-class assignments, activities, and exercises from active environmental communication instructors.

16 Writing a moral vision statement

Carrie P. Freeman

Students can get overwhelmed with all the negative information they learn about the state of environmental crisis (mass extinction, climate change, pollution, environmental racism, factory farming/fishing, water shortages, GMOs, dirty energy, etc.). Just as social movement communicators are encouraged to craft collective action frames around (1) diagnoses of the problem, (2) prognoses toward a solution, and (3) motivation for action (Snow and Benford 1988), so should environmental communication courses incorporate these components to inspire agency and efficacy for change. This "moral vision statement" assignment is inspired by linguist George Lakoff's (2004) recommendations to have activists articulate their moral vision for the kind of world they want to create, so they can explain what they are *for creating* in the long term and not just what they are against in the short term. Visualizing where they want to go in the end will help them articulate the kind of language and values to use in the initial campaigns to start society on a path to that sustainable vision. This encourages long-term thinking and creativity. Because the assignment is open-ended, instructors can expect variety in responses. I have found that some students write this assignment more about how they are changing their own lifestyles, and others write more broadly toward addressing the kind of global issues that most concern them—protecting ocean life, protecting human health, transitioning to clean energy, moving toward sustainable agriculture, etc. I choose to share my own moral vision statement with them as a model before they write theirs. Sharing is optional, but I do encourage instructors to write their own statement before they start to teach the course, so they gain the same insights the students will gain.

I have used this as the final assignment in every graduate and undergraduate environmental communication course I have taught. We end the class on a positive note and celebrate how we are each going to find our own path to creating a sustainable future. During the last class, I have the students bring in a hardcopy of their vision statement and do a "pair and share," where students swap statements, read, and discuss. This warms them up for a class-wide discussion of their ideas and visions.

'Moral vision statement' instructions for students

As this class hopes to be about change as much as evaluation/critique, we will each be encouraged to spend time in self-reflection at the end of the term to determine our values, beliefs, and vision for the kind of world we want to create that represents a healthy, sustainable way for humans to live. By the last class, each of you will bring a typed moral vision statement and also share your thoughts about it on that last day. This helps to end the term on a positive note where we don't just critique the status quo or discuss problems (and revel in the apocalyptic), but we inspire solutions and change strategies by defining a vision of what that viable society/ecosystem might look like and why it is preferred.

To provide an example for the format, I have posted my vision statement under "resources for moral vision statement homework." I hesitate to share my own vision because I don't want it to limit your own ideas, values, creativity, or areas of emphasis. Yours could be more or less utopian than mine, knowing that it is supposed to represent an ethical ideal—yet a vision for the future that you think makes sense and is not too fantastical. In mine, I chose to discuss how (and why) I envisioned humans relating ethically to (and valuing) other animals, plants, and ecosystems. Explore these types of question:

- What would that relationship look like, specifically?
- How would that work?
- What kind of world do you envision if you got everything you wanted—what is the end goal you are seeking by your advocacy efforts?

We have to know where we are going and what we want so we know how to get there (and we have to be *for* something good, not just *against* something bad). In that folder I also included the Earth Charter (2000) that might spark some ideas, as it represents multicultural values and guiding principles for a sustainable future. Additionally, Frances Moore Lappé's (2011) *Eco-Mind* book is also a great model and inspiration for envisioning a sustainable future and what that would look like.

Linguist and political strategist George Lakoff's (2004) book *Don't think of an elephant* recommends that activists articulate a vision of the world they want to create. Lakoff encourages people to talk in terms of values over facts and in the positive more so than the negative (we often find it easier to suggest what we don't like more than we can articulate what we do or would like—but this project is about doing the latter). He suggests that you carefully choose terminology that reflects your values and framings, rather than inadvertently reinforcing those of counter/oppositional or dominant groups. So the words you choose to use in your vision statement are important and telling. Choose them carefully.

The length should be in the region of 400–700 words, which is in the ballpark of one single-spaced page and is worth 5 percent of your grade. It is informal and personal in the sense that you don't have to include a reference page or cite anyone (unless you want to quote someone, then at least tell us whose words you are using; otherwise all words should be your own). But despite its brevity, and, in fact, because of its brevity (for a topic of such breadth), it will require some time to write and edit so that it is comprehensive, thoughtful, and useful. I think you will enjoy the process.

Keep a copy for yourself to put in your diary or a time capsule for safekeeping and historical documentation of your eco-brilliance.

Objectives and evaluation

The instructions include the objectives in terms of inspiring students to think positively and personally about their vision of a sustainable and just society so they know what they are working toward, can envision how this ideal society would work, and can articulate it to others. During the class discussion, they also benefit from learning what other students' visions are and hearing a diversity of ideas. An objective of mine is to end the course with a sense of optimism and hopefulness. In terms of student outcomes, I have been encouraged to see their enthusiasm and personal revelations on this assignment, and how much their interest in environmental and animal protection has grown over the course of the semester.

From an evaluation standpoint, I make it count enough that students will not skip it (e.g., worth 5 percent of their grade or a double-homework grade). I think this assignment should be graded largely for effort and completeness. For example, how well did they engage with the assignment to produce a creative and/or thoughtful response? I pay less attention to correcting grammar and spelling than I would on a research assignment or formal paper, although I expect the statement to show some sign of editing and proofreading so it does not read like a first draft. If it meets the word count requirements, follows instructions, and seems thoughtful and sincere, I generally give A's on this assignment, as it is the last one, and I see it mainly serving as encouragement for their future.

References

"The Earth Charter" (2000) Earth Charter Initiative. Available at: www.earthcharterinaction.org/content/pages/Read-the-Charter.html (accessed October 5, 2016).

Lakoff, G. (2004) *Don't think of an elephant: Know your values and frame the debate*, Chelsea Green Publishing, White River Junction.

Lappé, F. M. (2011) *EcoMind: Changing the way we think, to create the world we want*, Nation Books, New York.

Snow, D. and Benford, R. (1988) "Ideology, frame resonance, and participant mobilization," *International Social Movement Research*, 1(1): 197–217.

17 Environmental privilege walk

Unpacking the invisible knapsack

Tema Milstein and Stephen Griego

Environmental injustice and inequity can be abstract concepts for those who don't experience them and brutally real for those who do. The in-class environmental privilege walk activity helps bring course units that focus on these issues to life. This exercise, inspired by Peggy McIntosh's (1989) "White privilege: Unpacking the invisible knapsack," helps learners locate themselves within a spectrum of environmental privilege, centering the roles of socio-economic class, race, gender, sexuality, colonialism, and global disparities in environmental experience. The activity provides a foundation for understanding environmental injustice from the inside out, providing participants with a way to visualize privilege in their everyday experience. As participants take steps forward, privilege that formerly may have been invisible is made evident in unthreatening yet embodied ways. Participants have the opportunity to consider what this means in daily life for them, for others, and for the planet as a whole. Those who have experienced environmental marginalization or injustice are able to reveal discrepancies and often find their experience feels more legitimated and their voices heard. We hope increased levels of awareness and/or empowerment lead to higher levels of engagement in bringing about environmental justice.

To begin, participants stand shoulder to shoulder in an even starting line (either outside where there is space, or inside with desks pushed aside). As the instructor reads each statement, participants who silently answer "yes" take an equal step forward, those who silently answer "no" stay where they are. If space permits, alternately they can be instructed to take a step back. Many of the statements take some time to answer, as participants must thoughtfully consider layers of meaning. As privilege manifests itself through each step during the activity, the notion of a universal or equitable sameness in environmental experience is disrupted. In diverse learning groups, by the end of the activity difference in privilege and the roles of power and inequity in maintaining difference becomes spatially evident in the distances among participants.

Outcomes will vary. For instance, doing this activity with an indigenous community will differ from a large university classroom with a mixed population. Outcomes also are intersectional. Even with seemingly homogenous learning populations with presumed equal levels of privilege, those taking part may exhibit privilege differences related to gender, class, sexuality, etc. Most learners find they have experienced privilege in some way or on some level, especially those from highly industrialized countries. Tema originally created this activity for a United States southwest university setting, but context-specific statements can be included or substituted to make it suited to an array of educational settings and locations.[1]

Instructors can select the most effective statements to pose from the list below, depending on learner populations. Not all statements need to be posed. We have arranged statements from the personal/micro scale to the global/macro scale and from one's past to one's present. Below, we provide instructions for the activity, a list of statements to pose, a follow-up debrief with discussion prompts, optional supportive audiovisual artifacts, and suggested participant readings.

Activity: unpacking environmental privilege

Provide these instructions to participants:

A Please line up as if you are standing on a starting line, evenly shoulder-to-shoulder.
B I am going to pose multiple statements.
C If your answer is "yes" to a statement, step forward. If "no," stay where you are (or, alternately, take a step backward).
D (Instructors: Make sure you pause long enough for participants to meaningfully consider each statement. Encourage them to ask for clarification. Participant questions during the activity can lead to deeper understandings.)

- I grew up in a place where I felt safe and had plenty of outdoor areas to play in.
- The environment I grew up in had clean air, land, and water.
- Now, I daily have access to healthy filtered water.
- I have access to fresh, local, and affordable organic produce.
- Where I live, I can access a quiet and safe outdoor place, day or night.
- I can choose to live in a neighborhood where many local businesses are accessible by walking or bicycling.
- If I wish to reduce my carbon footprint, I can afford to purchase a hybrid or electric vehicle.
- I can nap in a public park without it being assumed I might be homeless.
- I can work in my own yard or garden without passersby assuming I am hired help.
- If I spend time in wild places, I encounter people who look like me.
- If I spend time alone in wild places, I feel welcome and safe when I encounter other people.
- I have mostly positive associations with rural locations.

- When people refer to me as in some way natural or animal-like I assume it is a compliment and not a put-down.
- I can spend time outdoors in border zones without authorities stopping me.
- I can choose to spend time outdoors only when the weather is agreeable.
- I can enjoy nature preserves or national parks, imagining them as intact wildernesses because their establishment did not involve the forcible removal of my ancestors.
- I have traveled to and experienced a wide variety of nature in different parts of the world.
- I can think about climate disruption as an abstract concept because I do not experience my own region or my own survival as immediately threatened.
- I can consume whatever material goods I want when I want without noticing obvious negative effects on my land, my water, my country, etc.
- As climate disruption begins to negatively affect the place I live, I have the option to move.

Activity debrief

Discussion Prompts

After the final privilege statement is posed, while participants are still in their final locations, ask them the following:

- Look around: Notice where are you in relation to where you started and in relation to others.
- There are three things I want you to notice: the distance you covered or did not cover, your proximity to others, and the clusters of people.
- Is this where you thought you would be on the spectrum?
- Why or why not?
- Do you see any patterns or discrepancies?

Move to a discussion circle and ask the following:

- What did this exercise bring up for you?
- Do you have a salient memory that related to any of the statements I posed?
- What did the statement reveal about socio-economic class, gender, race or ethnicity, sexuality, national citizenry, age, etc.
- If there was an uneven (or even) distribution of privilege, do you see this as a problem?
- Should any of these privileges be the norm for everyone? Should any not? Why?
- How might not having some of these privileges affect your life experience?
- How might having some of these privileges affect your life experience?

- How might having more privileges than others possibly distort your view of others' environmental experience?
- Looking forward, how might this activity inform your thoughts or actions?
- Those of you who walked farthest in this activity, now that you are aware of this privilege difference, what will you do with it? How might this activity help you to discuss or work for environmental justice?
- Those of you who made the least forward movement, how might this activity inform ways you discuss or work for environmental justice?

(Teaching note: In this activity, those at the front are often most surprised and those at back least. The kind of privilege addressed in this activity is generally systemic, unearned, and, to those who have it, invisible. In addition, the most environmentally marginalized often are unheard and unseen in this activity as they are not participants; so, one can ask participants how the activity might have differed if the learning space were expanded to include representatives of all people.)

Optional Supportive Audiovisual Artifacts

Tema follows up the activity by asking participants to define environmental injustice and environmental justice. After some discussion, regionally relevant examples can help expand the conversation. Some examples in the United States include:

- University of California San Diego student Bryant Pena's (2009) spoken word class project on environmental racism and First Nations people that helps center indigenous experience (Tema plays 1:59–5:06 minutes).
- Ron Finley's (2013) TED talk on guerilla gardening and South Central Los Angeles that helps students consider whether environmental justice can start at the individual scale from within communities.
- Majora Carter (2007) speaking about her work in New York or, alternatively, the work of Greencorps Chicago (2008) share ways people are organizing to overcome environmental injustice.

Note

1 For example, Mengel's (n.d.) "Race and Class Privilege in the Environmental Movement" statement list, which helped inform statements in the exercise, has been used by the Pachamama Alliance in its trainings.

References

Carter, M. (2007) "Greening the ghetto" [Ted Talk video]. Available at: www.youtube.com/watch?v=gQ-cZRmHfs4 (accessed October 4, 2016).

Finley, R. (2013) "Guerilla gardening in South Central Los Angeles" [Ted Talk video]. Available at: www.youtube.com/watch?v=EzZzZ_qpZ4w (accessed October 4, 2016).

Greencorps Chicago (2008) "Stories from the Green Collar Economy" [Ted Talk video]. Available at: www.youtube.com/watch?v=izhPxPjsnvA&feature=channel (accessed October 4, 2016).
McIntosh, P. (1989) "White privilege: Unpacking the invisible knapsack," *Peace & Freedom*, July/August, 10–12.
Mengel, G. (n.d.) "Protecting the source, inspiring the future," Pachamama Alliance. Available at: www.pachamama.org/news/race-and-class-privilege-in-the-environmental-movement (accessed June 10, 2015).
Pena, B. (2009) "Native environmental racism." Available at: www.youtube.com/watch?v=rdKGNwQXvoU (accessed October 4, 2016).

Further reading

The following readings, read by participants before the exercise, have been useful over the years. Some are more suited to graduate courses and some to undergraduate courses.

Bell, M. (2004) "Body and Health," in Bell, M., *An Invitation to Environmental Sociology*, Sage Publications, Thousand Oaks, pp. 111–31.
Broad, G. (2013) "Ritual communication and use value: The South Central Farm and the political economy of place," *Communication, Culture & Critique*, 6(1): 20–40.
Carmin, J. and Agyman, J. (2011) *Environmental inequalities beyond borders: Local perspectives on global injustices*, MIT Press, Cambridge, MA.
Escobar, A. (1999) "After nature: Steps to an antiessentialist political ecology," *Current Anthropology*, 40(1): 1–30.
Endres, D. (2009) "The rhetoric of nuclear colonialism: Rhetorical exclusion of American Indian arguments in the Yucca Mountain nuclear waste siting decision," *Communication and Critical/Cultural Studies*, 6(1): 39–60.
Evans, M (2002) "'Nature' and environmental justice," in Adamson, J., Evans, M. M. and Stein, R., eds., *The environmental justice reader: Politics, poetics, and pedagogy*, University of Arizona Press, Tucson, pp. 181–93.
Milstein, T. and Dickinson, E. A. (2012) "Gynocentric greenwashing: The discursive gendering of nature," *Communication, Culture, & Critique*, 5(4): 510–32.
Mortimer-Sandilands, C. and Erickson, B., eds., (2010) *Queer ecologies: Sex, nature, politics, desire*, Indiana University Press, Bloomington.
Pezzullo, P. (2001) "Performing critical interruptions: Stories, rhetorical invention, and the Environmental Justice movement," *Western Journal of Communication*, 65(1): 1–25.
Plumwood, V. (1997) "Androcentrism and anthropocentricism: Parallels and politics," in Warren, K. J., ed., *Ecofeminism: Women, culture, nature*, Indiana University Press, Bloomington, pp. 327–55.
Taylor, D. (2000) "The rise of the environmental justice paradigm: Injustice framing and the social construction of environmental discourses," *American Behavioral Scientist*, 43(4) 508–80.

18 An experiential approach to environmental communication

Emily Plec

The Environmental Communication (EC) classroom is a space of personal transformation and positive environmental change. I am among those environmental communication scholars who believe we "have an ethical duty to not only try to explain but also help change the society that has caused ecological collapse ..." (Milstein 2009, 348). This belief informs my curriculum and the progressive series of experiential learning activities described below. To be clear, the society to which I refer is one marked by late consumer capitalism and relative privilege. Though the students I teach are mostly working class and many are the first in their families to attend college, they exist within that social formation of US consumerism with varying degrees.

In my introductory EC course, I begin with several definitions of environmental communication, including Peterson's (1998, 385) assertion that EC "is about symbolic practices for constituting knowledge, relationship, and identity," and Hendry's (2010, 3) argument that the field "seeks to describe and understand the role that communication plays in establishing the meaning of nature and humans' relationship to the natural world, and to influence human action as it pertains to the natural environment." We discuss the material and symbolic dimensions of EC, emphasizing ways language and representation influence the material world. For, as Milstein (2009, 346) points out, this "brings power to the forefront" and demonstrates ways "cultural, economic, and ideological forces inform representations of nature, constraining or allowing for particular ways of communicating about the environment." These definitions provide the framing for the activities included here. I then issue a fair warning to the class:

> For this course, I require you to do three difficult things (in addition to other, less strenuous, assignments). First, I ask you to consider your relationship to the natural world in terms of environmental impact by tracking your consumption of resources for one week. Second, I expect you to try to live for one day without many of the environmentally degrading products and practices to which many of us have become accustomed. Finally, the course requires that you put communication principles into action by working collectively with others on an environmental service project.

Taken together, these endeavors produce a set of learning outcomes that are consistently met by students who make a sincere effort to participate in them. These outcomes include the ability to:

- examine their own implication in environmental issues;
- reflect upon the role communication plays in shaping their environmental choices, impacts, and actions; and
- practice applied environmental communication in concert with others.

In the sections that follow, I provide instructions for each of the exercises, describe how each is conducted and evaluated, and discuss ways the outcomes are manifested in students' responses to the assignments.

Exercise 1: tracking consumption

This assignment requires you to consider your impact upon the environment in terms of your consumption of resources and material goods. For six days, keep a careful and detailed log of your consumption habits in all categories. Double check the list each night before you go to sleep to make sure you have recorded all of the day's activities. Be as detailed as possible when keeping your records and update frequently! Take note of both patterns and their exceptions. For example:

- At what temperature do you keep your home and for how many hours per day (heat/AC)?
- How long does the water run while you brush your teeth or shower?
- How many times do you flush the toilet each day?
- How many hours is the TV or Computer on?
- How often do you use and charge your phone and for how long? What foods/drinks do you prepare & how?
- What goes into your trash each day?
- What gets recycled? Composted?

The list below will aid you in tracking your consumption over the next six days:

Gas
Natural gas (home heating/stove/water/grill)
Automobile fuel

Electricity and Batteries
Lights
Electric heat (home/stove/water)
Appliances and electronics (including those drawing electricity 24 hours a day)
Other electrical or battery powered devices (TV, Computer, music player, phone, game system, car battery, clock, etc.)

Water
Personal hygiene (washing hands, brushing teeth, bathing/showering, flushing toilet, etc.)
Cooking and cleaning
Drinking water
Indoor/Outdoor watering (plants, sprinklers, lawns, etc.)

Other Consumables
List processed and/or packaged food and drinks vs. bulk, unprocessed and/or fresh foods
Cleaning products (personal hygiene and home)
All product purchases (groceries, clothes, school supplies, entertainment, snacks, etc.)
Make a note of everything you throw in the trash (or recycle or compost) each day.

Writing up the exercise

Take some time to review your notes from the past six days. Answer the following questions in a brief two-page reflection paper and attach your detailed tracking log:

1. What patterns of consumption appear over the six-day period for each of the categories you tracked? Were you surprised by any of the results? Explain.
2. What sorts of things would be hard to give up or change? Which items do you depend on most? What was most difficult about the assignment?
3. Reviewing each category of consumption above, what could you do now and in the future to reduce your consumption of natural resources without negatively affecting your quality of life? Set a few reasonable, specific, and concrete goals for each section and explain how you will go about achieving them.

Discussion

The writing assignment is followed by "environmental footprint" calculations (www.footprint.org/calculator) to enable further reflection on the planetary impacts of factors such as housing, food, employment, and transportation. Our class discussions highlight the challenges of the exercise and what societal, economic, political, and cultural changes are needed to significantly reduce our footprints. The students then discuss, in small groups, the role of communication in those changes, and together we draw linkages to mass media, political and organizational communication, rhetoric, and interpersonal communication.

The exercise is evaluated on the completion of the written reflection paper and the level of detail in the tracking log. Students who fail to address all questions in the above assignment instructions and/or omit categories of resource consumption are docked points for the omissions. No judgment is rendered regarding the amount of resources consumed nor the goals students set, as the point is not to criticize individual behaviors but to heighten awareness of our implication in environmental issues and reflect upon the role communication plays in constructing lifestyles that are more or less environmentally sustainable. Evidence of these outcomes is manifested in students' responses to the exercise, including awareness of the role of family or roommate behaviors, lack of control over housing and transportation infrastructure, and socio-cultural ideas about comfort and consumerism.

Indeed, the combination of tracking consumption and calculating environmental footprints is effective for these students, in part, because of the socio-cultural location they occupy. Although heterogeneous in most cultural and socio-economic respects, college students in many countries occupy a position of relative privilege to others, both in their larger society and across the globe. This dynamic of relative privilege is further highlighted by those students who attribute relatively small footprints (2–3 earths in my western U.S. classroom) to economic hardship; mass transit, dumpster diving, community gardens, and efficiency apartments are not environmental choices so much as economic necessities.

Exercise 2: low impact day

Now that you are aware of the natural resource requirements your life entails, you have an opportunity to experience living your life with minimal environmental impact. For this "Low Impact" exercise, you must examine and suspend what may have become habits related to gas, electricity, batteries, transportation, food, and drink. For one day (from the time you wake up until you go to bed), you should avoid using gas, electricity, batteries, or a private vehicle. You are permitted to use non-vehicular transportation (bicycle, chair, skateboard) or mass transit. You may not consume processed foods, drinks, or bottled water (unless required for safety) and should avoid packaging unless you recycle or compost it. Try not to consume any foods that come from more than 200 miles away and make a note if you do. Finally, for this one day, you must fulfill any social needs without spending money (you are permitted to use money only to purchase mass transit tickets or fresh, local foods). If you break one of the rules, explain which rule you broke, why you broke it, and what would have to happen for you to be able to follow it.

Write up your journal about your experience throughout the day. Include your reactions to each facet of the assignment and any difficulties and challenges you experienced:

- What were the highlights?
- What did you do differently during this day?
- What did you notice?
- What enduring lessons might come from your "Low Impact" Day?

The assignment will be graded based on the extent to which you endeavor to follow the rules as well as the completeness of your journal and responses to the above questions.

Discussion

This second activity is announced early in the term and is due toward the end. Students are encouraged to schedule their "Low Impact" day in advance to increase the likelihood that they can observe all of the rules. The journal is evaluated according to the above instructions and the depth of students' responses to the questions posed. The exercise due date is followed by an in-class viewing of Gabbert and Schein's (2009) documentary film *No Impact Man* with discussion and writing prompts that encourage students to reflect and compare their experiences to those of the family profiled in the film. The discussion also emphasizes the role mass media representations play in shaping attitudes and assumptions about humans' environmental relationships.

The goals of the exercise are to force students to examine and assess their needs and wants, and consider how a culture of unsustainable consumption is constructed (and can be contested) through communication. The exercise is stressful for some students, yet, in their write-ups, many report additional positive outcomes as a result of the exercise, including increased quality of interpersonal interactions, more time to sleep and study, feeling healthier (though many also report being hungrier), and spending more time outdoors than usual. As Peterson (1998, 372) notes, EC research "should suggest how to utilize the human capacity for language to understand and reflect upon technologized conditions of our own making, but not necessarily to our liking." As a class, we discuss how technology can facilitate and hinder conditions of our liking, as well as what other kinds of social, economic and cultural formations are needed to enable sustainable living.

Students share ideas about how the demands of the activity could be more easily met through frequent and efficient mass transit, affordable localized food systems, green building and housing, and exercising greater control over communication technologies such as email, texting, and social media. Asked, "What would it take?," students are charged with searching for answers and finding out who is proposing them. We then, in a subsequent class session on environmental advocacy, examine the rhetoric of activist groups, NGOs, and government agencies identified by students in their searches.

Final service project

The final project moves us from an awareness of our implication in, and impact upon, the environment to deliberate engagement with others in a collective environmental effort. At the beginning of the term, students choose a service project option (or propose their own) and begin working collaboratively on the project with other students and/or with community partners for a total of roughly fifteen hours. The purpose of the service project is to demonstrate the limitations of personal actions/responsibility as a stopping point (or the only means) for environmental action and to empower students to join with others to effect positive environmental awareness and change. It is important that students understand they need not be perfect environmentalists to engage in critical inquiry and action. The service project is evaluated according to four equally weighted component parts:

- a proposal in which students conduct initial research and set goals for their project;
- an external service partner evaluation of their contribution to the effort;
- a short self-reflection essay; and
- a final class presentation in which they highlight the EC components of their projects.

Typical project options include: working with local environmental groups on campaigns, volunteering with the campus garden, conducting workplace environmental assessments, and engaging in conservation group projects.

The combination of these three experiential learning activities has been highly effective in encouraging students to be more conscious, aware, and considerate of their environmental relationships and impacts while also recognizing the potential for substantial and impactful change through collective action. Together, these three activities aid in deconstructing ideas about the environment as wilderness, or "the big out there" (Cronon 1995, 86). They also result in students identifying themselves as environmentalists and their practices as environmental actions that have real, material consequences. Moreover, students internalize an understanding of EC as a means for creating knowledge, relationships, identities, and action.

References

Cronon, W. (1995) "The trouble with wilderness; or, getting back to the wrong nature," in Cronon, W., ed., *Uncommon ground: Toward reinventing nature*, Norton, New York, pp. 69–90.

Gabbert, L. and Schein, J. (directors) (2009) "No impact man" [DVD], Oscilloscope Laboratories, Brooklyn, NY.

Hendry, J. (2010) *Communication and the natural world*, Strata Publishing, State College, PA.

Milstein, T. (2009) "Environmental Communication," in Littlejohn, S. W. and Foss, K. A., eds., *Encyclopedia of Communication Theory*, Vol. 1, Sage Publications, Los Angeles, pp. 341–9.

Peterson, T. R. (1998) "Environmental communication: Tales of life on earth," *Quarterly Journal of Speech*, 84: 371–93.

19 Greening epideictic speech

Jake Dionne

"Greening Epideictic Speech" is a unit-length activity that tasks students with developing and delivering an original ceremonial address that shows consideration of the "more-than-human world" (Abram 1996, 7). In this activity, students *green* epideictic rhetoric (a speech genre of praise or blame) by creating and executing either *a eulogy about an extinct species of nonhuman* (e.g., Dodo Bird, Tasmanian Tiger, or Great Auk), or *a speech of inspiration to a community affected by a natural disaster* (e.g., New Orleans, Louisiana after Hurricane Katrina; Fukushima, Japan after the 2011 earthquake; or Fort McMurray, Alberta after the 2016 wildfire). By infusing epideictic rhetoric with environmental *topoi*, this activity instructs students about an oft-delivered genre of speech all whilst responding to the posthuman call to showcase humanity's entanglement with the more-than-human world.

It would be a cruel exaggeration to suggest that this activity serves as a corrective for the longstanding tradition of anthropocentric communicative practices that privilege the human inhabitants of the more-than-human world (Goodale and Black 2010); however, it has been my experience that this activity does suspend—even if only briefly—the deep-rooted strand of "human exceptionalism" that tangles with students of the humanities (Worsham 2013, 53). This activity does so by creating the conditions for students to experience a form of empathetic unsettlement with how their species communicates about the more-than-human world. That is, this activity asks students to consider themselves *always already* interconnected with the more-than-human world by forcing them to undergo an affective "kind of virtual experience through which one puts oneself in the other's position while recognizing the differences of that position and hence not taking the other's place" (LaCapra 2001, 78). Consequently, this activity transforms the epideictic speech into an ecopedagogical tool, a change that aligns with what I described elsewhere as the need to ecologize traditional building block assignments in the basic communication course (Dionne 2016).

Assignment walkthrough

Designed for introductory-level courses in presentational speaking, two underlying presumptions govern "Greening Epideictic Speech":

- First and foremost, this activity presumes that students do not know how to design and perform an organized epideictic speech. Therefore, instructors are responsible for charting the theoretical underpinnings of this type of rhetoric in order to familiarize students with the characteristics of epideictic speeches, a genre of oration categorized by Aristotle himself.
- Second, this activity presumes that students are unfamiliar with discussions of extinction and natural disasters as presented in the literatures of critical animal studies and environmental studies. As such, instructors are responsible for introducing students to generative and critical discussions about humanity's entanglement with the more-than-human world.

By way of introduction, students must first read a select number of texts that will orient them to ongoing conversations regarding both epideictic rhetoric and the more-than-human world. These texts include the following articles and chapters:

- "The functions of epideictic: The Boston Massacre orations as exemplar" (Condit 1985);
- "Embracing humanimality: Deconstructing the human/animal dichotomy" (Freeman 2010); and
- introduction to *The ecological thought* (Morton 2010).

I also encourage instructors to utilize a textbook that includes a section on epideictic speeches (e.g., O'Hair et al. 2015). Taken together, these readings will help students build a critical understanding and vocabulary about the uses of epideictic rhetoric as well as the human/animal and nature/culture dualisms, two anthropocentric communicative phenomena that work to conceal humanity's entanglement with the more-than-human world.

After completing these readings, evenly separate students into two groups named *Eulogizers* and *Inspirers*. Present the Eulogizers with the topic: *Design and deliver a eulogy for an extinct species of nonhuman*. Present the Inspirers with the topic: *Design and deliver a speech of inspiration to a community affected by a natural disaster*. Each individual student ought to research, compose, and extemporaneously deliver a five- to seven-minute oratory related to his or her assigned topic.

The next step in this activity requires that students further refine the scope of their speech, a process that grants students agency to determine what species of nonhuman or instances of natural disaster they want to research and discuss. Depending on the available modes of communication (e.g., Blackboard, Desire2Learn, e-mail, etc.), instruct students to submit a *subtopic selection form*. This form should include space for students to request a particular species of nonhuman or community affected by a natural disaster as well as to justify their choice in a 250-word rationale that documents their background knowledge about the topic at hand. The goal of this subtopic selection form is to provide students with a platform to brainstorm potential pitfalls and possibilities of their epideictic speech in addition to preventing students from selecting the same subtopic. Because breadth of content is desirable for cultivating a broad understanding of nonhumans and natural disasters, it is important to bar students from selecting the same subtopic.

The Eulogizers select a species of extinct nonhuman (e.g., West African Black Rhinoceros, Quagga, Golden Toad). Do not allow Eulogizers to select a species of endangered nonhuman, because many of these presenters will focus on what humans can do to further prevent biodiversity loss. While attention to conservation efforts is productive, Eulogizers should focus exclusively on loss and how the orator and audience can come together to form a compassionate, empathetic community that memorializes and celebrates nonhumans rather than a legislative public.

The Inspirers select a specific community affected by a natural disaster (e.g., Bridge City, Texas after Hurricane Ike; Valdivia, Chile after the 1960 earthquake; or Karachi, Pakistan after the 2015 heat wave). Encourage a number of Inspirers to consider international communities and historic natural disasters (e.g., Boston, Massachusetts, after the Great Blizzard of 1888; Pompeii, Italy, after the eruption of Mount Vesuvius in AD 79; or Helike, Greece, after the tsunami and earthquake of AD 373). By selecting geographically and temporally diverse topics, instructors and students create the conditions for interesting and expansive dialogues about humanity's entanglement with the more-than-human world on a deeper historical level.

After garnering approval for their subtopics, students must then construct outlines for their epideictic speeches. Students should approach this document by writing a full sentence outline, a speaking tool that typically spans five to seven double-spaced pages of text. When discussing the ways in which students might orchestrate their eulogies or species of inspiration, urge them to consider the myriads of constraints and opportunities associated with their subtopic.

Eulogizers must not fetishize the strife experienced by nonhumans. Two points of consideration are crucial to avoid trivializing extinction:

- Eulogizers should focus on "celebrating and commemorating the life" lived by nonhumans rather than ascribing blame to those humans who survived (O'Hair et al. 2015, 396); and
- Eulogizers should include only facts and narratives that best illustrate the unique embodied experience of the selected nonhumans. If the goal of the presentation is the communal embodiment of empathetic unsettlement, then students must avoid speaking "in terms of uplifting messages or optimistic, self-serving scenarios" through which students act as saviors and pretend as if they personally knew their selected nonhumans (LaCapra 2001, 78).

Likewise, Inspirers should use caution when deploying the rhetoric of blame in their epideictic speeches. Whereas blame is appropriate for this genre of speech, rhetors often haphazardly indict natural disasters themselves as culprits for communal collapse and despair when, in fact, these forces of nature merely create the conditions for communicative hierarchies to crumble and are reborn anew. Take, for example, Hurricane Katrina. Although this natural disaster flooded New Orleans, Louisiana, it was not the storm itself that caused racial unrest in the community. Rather, the storm disrupted a community already ravished by racism and spurred an unraveling of tightly woven post-racial narratives that dominated the American cultural imaginary about the South. Accordingly, a successful speech of inspiration might locate racism as a disparaging communal value, indicate how the natural disaster created conditions that exposed such a flaw, and declare a set of values that the community might adopt in moving forward that will better aid citizens in both responding to the tragedy at hand and living with less systemic violence.

Beyond the aforementioned components, instructors may choose to include supplemental assignments relevant to their individual courses. In my course on presentational speaking, I encourage students to include peer-reviewed sources in their outlines. I do not, however, require students to verbally cite their sources, as this stylistic convention often works against students' capacity to speak with creative prose. What I do require is that students include visuals (e.g., PowerPoint or Prezi). Whereas other speech genres warrant visuals like charts and graphs, I limit the number of slides students can include to one impactful image. In short, creative additions should be modified to fit each individual course and a classroom's technological ability.

As an outcome of this activity, students should be able to perform the following four tasks:

- identify the key characteristics of epideictic rhetoric and what differentiates this speech genre from judicial and deliberative orations;
- compose either a eulogy or speech of inspiration that engenders communal empathy about both human and nonhuman bodies affected by extinctions and natural disasters;
- locate the logics of anthropocentrism as they permeate through larger collections of discourses; and
- recognize humanity's entanglement with the more-than-human world.

In evaluating students, instructors must accept the inevitability of anthropocentric behaviors and attitudes. As a system of power, anthropocentrism cannot and will not be eradicated in the basic communication course. Given that both epideictic rhetoric and discussions of the more-than-human world are unfamiliar to students, instructors must use anthropocentric happenings as productive moments of conversations. For example, if Eulogizers pretend to cry, applaud students for their effort in attempting to emote their eulogies while simultaneously using these fetishistic moments as a catalyst to discuss the ways in which fake tears might demonstrate a lack of consideration for nonhumans. Likewise, if Inspirers blame their natural disasters instead of locating dangerous systems of power at play in particular geographical locations, commend students for exploring the more-than-human world in addition to reframing the conversation to include the ways in which humanity creates the conditions for disaster. On the whole, assign positive marks to students who demonstrate *an attempt* to establish an empathetic connection between the human and more-than-human world and approach failures as learning opportunities.

In assigning a grade and offering feedback, instructors should formulate a rubric that takes into equal consideration the form of epideictic speech and the ecological content thereof. On this rubric, include not only procedural elements (e.g., subtopic selection form, outline, and time length), but also the *performance of empathy*. In other words, ask the following question:

- What formal and/or stylistic maneuvers did the student use to enact the role of an empathetic rhetor conscious of humanity's entanglement with the more-than-human world?

To earn full credit, a student must show a degree of reflexivity in regard to their subject position.

Since both eulogies and speeches of inspiration require that students step outside their comfort zones and consider topics beyond their usual repertoires, it has been my experience that students undergo a form of embodied cognitive dissonance and subsequently walk away from this assignment with a new appreciation for ecological concerns. Indeed, students have reported this cognitive dissonance in their evaluations of this activity, citing, for example, how "it feels weird to memorialize animals that we [human students] did not personally know."

In conclusion, "Greening Epideictic Speech" addresses an exigent need in the humanities, especially rhetorical studies, to discuss the more-than-human world by transforming the traditional epideictic speech into an ecopedagogical tool. "The eulogy project made me appreciate the voices of those fighting against extinction," noted one student in her evaluation of this assignment, "Having everyone pick a different animal helped show the far reaching effects of extinction while paying respect to those who have already fallen." Another student wrote, "Upon writing my speech of inspiration for African-American communities of New Orleans, my biggest takeaway was how Hurricane Katrina, although fatal, helped expose racism in the community." If we view the presentational speaking classroom as a place for students to rehearse what it means to be a responsible human in a more-than-human world, then this activity succeeds at raising consciousness about several of the woes that define the geological moment many of us are calling the Anthropocene.

References

Abram, D. (1996) *The spell of the sensuous: Perception and language in a more-than-human world*, Vintage Books, New York.

Condit, C. M. (1985) "The functions of epideictic: The Boston Massacre orations as exemplar," *Communication Quarterly*, 33: 284–98.

Dionne, T. J. (2016) "Mapping human sacrifice zones through informative speeches," *Kaleidoscope*, 15: 61–7.

Freeman, C. P. (2010) "Embracing humanimality: Deconstructing the human/animal dichotomy," in Goodale, G. and Black, J. E., eds., *Arguments about animal ethics*, Lexington Books, Lanham, pp. 11–30.

Goodale, G. and Black, J. E. (2010) "Rhetoric and 'animals': A long history and short introduction," in Goodale, G. and Black, J. E., eds., *Arguments about animal ethics*, Lexington Books, Lanham, pp. 1–7.

LaCapra, D. (2001) *Writing history, writing trauma*, The John Hopkins University Press, Baltimore.

Morton, T. (2010) *The ecological thought*, Harvard University Press, Cambridge, MA.

O'Hair, D., Stewart, R. and Rubenstein, H. (2015) *A speaker's guidebook: Text and reference*, Bedford/St. Martin's Press, Boston.

Worsham, L. (2013) "Toward an understanding of human violence: Cultural studies, animal studies, and the promise of posthumanism," *Review of Education, Pedagogy, and Cultural Studies*, 35: 51–76.

20 Praxis-based environmental communication training

Innovative activities for problem-solving

Bridie McGreavy, Caroline Gottschalk Druschke, Leah Sprain, Jessica L. Thompson, and Laura Lindenfeld

Environmental communication (EC) focuses on understanding and solving complex and urgent social and environmental problems (Cox 2007; Lindenfeld et al. 2012; Sprain and Timpson 2012; Endres et al. 2008). For us, this has meant finding ways to bring EC to the fore of both classroom teaching and interdisciplinary, cross-institutional partnerships (McGreavy et al. 2016). We came together to write this chapter because each of us has been working for more than a decade in classroom and community-based contexts where problems related to climate change, land use and water quality, poverty, and environmental justice have pressed in on and profoundly shaped our teaching and research. As students in our classes learned about the impact of dams on human communities and migratory fish, for example, they wanted to know how they could get involved in addressing these problems in meaningful ways. And, as community partners struggled to make sense of rapid shifts in their home ecosystems from climate change, they wanted to find better ways to work together and bring diverse perspectives to understanding the changes and finding ways to adapt. In these and many other instances, EC praxis has helped us create learning opportunities for students and collaborators to develop the conceptual and practical tools to solve complex problems together.

This set of activities builds interpersonal and critical reflection skills for inclusive, effective, and creative approaches to solving pressing sustainability problems (see Table 20.1). Though the activities can be used individually, we put them in a sequence starting with skill development to help people open up to differences in perspectives (Activity 1); discuss and negotiate those differences (Activity 2); critically reflect on the process of problem identification and negotiation (Activity 3); begin to knit the differences together into shared understanding (Activity 4); and work together on a case to implement these skills (Activity 5). Through EC, we know there are multiple ways to understand any given problem and often multiple possible solutions, too. When we respond to specific crises and problems, we are also mindful that communication actively shapes what we come to see as a problem and our judgments about how to act (Schwarze 2007). Thus, it is essential that in collaborative problem-oriented

Table 20.1 Framework to orient the design and assessment of activities that build specific capacities for problem-solving.

	(1) Yes, and . . .	*(2) Problem framing*	*(3) Mirror, mirror*	*(4) Shared systems mapping*	*(5) Collaborating on a case*
Group size	25–50	10–20	6–20	6–12	20
Ideal timing during a collaborative effort	early	early to mid	mid to end	early to mid	any
Context and roles					
Classrooms	X	X	X	X	X
Research teams	X	X	X	X	
Interested citizens	X	X		X	X
Community organizations	X	X	X	X	X
Businesses	X	X		X	
Governmental (elected and appointed, across scales)		X	X	X	
EC praxis for problem solving					
Adaptive capacities that focus on accepting complexity, ambiguity, change	X	X			X
Active listening and openness	X		X		X
Framing and symbolic construction		X		X	X
Critical reflection	X	X	X	X	X
Interpersonal communication and interactional expertise	X	X	X		X
Systems thinking		X		X	X

Source: adapted from McGreavy et al. (2016)

Note: The group sizes, timing, contexts, and praxis commitments reflect how we have used these activities, and we encourage active experimentation for new applications and skill development.

projects people learn how to approach an issue from multiple perspectives and create spaces where deliberation leads to new understanding, insights, and progress toward identifying and implementing a range of solutions.

Activity 1: "Yes, and ..."

Trying to solve problems through classroom and community-based collaborations requires a commitment to remain open and to learn and change together. This activity takes Bakhtin's (1990, 933) notion to heart, where "a word is a bridge, thrown by myself and another. If one end of the bridge depends on me, then the other depends on my addressee." "Yes, and ..." is a bridge-building activity that teaches people how to stay open to multiple perspectives and find ways to connect disparate words and ways of thinking about a problem. We have found this activity is well suited for situations where there are many different perspectives that need to be brought together for some kind of shared decision-making, such as developing an interdisciplinary research program to inform policy, identifying goals for natural resource management planning, or helping students plan group projects.

"Yes, and ..." borrows an improvisational theater technique of affirming and building on another speaker's contribution to help teams in the early stages of designing a joint project. A facilitator divides participants into small groups with one member identified as scribe to track key ideas. The facilitator prompts groups to focus on their project plans by asking them to write down their initial ideas or discuss in a small group to do some early brainstorming. The activity commences with one individual responding to the prompt while others listen. The next individual must "improvise" by acknowledging what he or she has just heard and describing how his or her research might intersect with the idea previously presented. The process moves forward until all members in the group have contributed. The willingness to listen and consider how each person contributes to a greater whole is critical to this exercise. At the end, the facilitator focuses discussion on the ideas generated within the groups and on the process itself. This encourages individuals to talk about how it feels to work in a different mode and self-reflect on their abilities to listen, engage, and create new collaboratively developed ideas. Author Scott McDowell (n.d. para. 2) describes this activity as a "volley back with something your fellow players can respond to," emphasizing how "Yes, and ..." promotes collaborative thinking, prioritizes openness, and conceptualizes the possibility in every moment.

This exercise is especially effective in generating a bigger picture of an issue, problem, context, or opportunity because it pushes individuals to create a genuine connection in a playful yet empathic way. When we have used this activity with interdisciplinary teams, it has enabled people to configure new ideas for papers, presentations, and grant proposals that draw on intersections generated by the exercise itself. New ideas have emerged that sit above the ideas of any one individual, and these ideas have led to new, unexpected collaborations. Using this activity in a class can help students develop group projects as it gives students a specific strategy for inviting all members of the group to contribute their ideas in an open and inclusive way.

Activity 2: problem framing

Because problems are inherently multi-dimensional with many different ways of defining what a problem is and who that problem affects, developing skills in problem framing is an important focal area for EC (Sprain 2013). This activity helps people enact a "problem-posing, problem-solving mode of inquiry" engaging in judgment and invention to learn how to weigh alternatives and decide the kind of future we want to make together (Schwarze 2007, 97). The problems we address in EC can be considered wicked ones (Kreueter et al. 2004, McGreavy et al. 2013). Wicked problems are characterized by inherent complexity; multiple and sometimes conflicting perspectives and values; social and ecological issues that are deeply entangled and mutually influencing; and the potential for unintended consequences, as when fixing one part of the problem ends up creating a new problem. Given these challenges, a problem-oriented mode of inquiry can be organized in stages to help collaborators take one step at a time as they identify the problem, negotiate differences in understanding it, and then create new ways of thinking about and solving problems.

The Problem Framing activity is designed to help students and collaborators develop skills in the early stages of problem definition to identify and work through the frame differences that shape how collaborators come to understand the issue at hand (Dewulf et al. 2007). Though this activity is probably more useful in the earlier stages of a collaboration, it can also be used any time when returning to problem definitions would help participants understand differences and forge new ways of understanding the situation they are trying to address. This exercise invites people to consider how problem definition is a frame and how frame differences shape collaborations. Participants begin by writing for about fifteen minutes in response to one or more of the following prompts:

- What is the problem you are working on?
- Who (or what) has experienced this problem?
- How did you come to see this as a problem?
- How do you approach this problem?
- In what ways do you think you can address this problem?
- What progress have you made thus far?

Participants then pair up and exchange stories. They are asked to listen with intent and remain quiet as their partner speaks. Throughout the listening cycle, participants are asked to think about how they might approach the partner's problem in a similar or different way. After the listening rounds, the pairs discuss how they can work together to address this problem. In the final discussion phase, the group as a whole explores questions such as:

- How did your understanding of your problem change through this dialogue?
- What kinds of interactions helped you to understand the problem from this new perspective?
- What didn't work and how could you as individuals and as partners have approached the dialogue differently?

We have used this activity in diverse classroom and community settings. For example, while waiting for a revised Environmental Impact Statement on a proposed reservoir, community leaders wanted to begin a community conversation about water issues without focusing on the polarizing reservoir. An interdisciplinary graduate class on water conflict engaged in problem framing for how the community would meet future water needs, where each student framed the problem in public terms and identified three or four different approaches to address the problem. Conversation about these framings opened up different ways of understanding water conflict and provided the basis for a public framing for deliberation around the question:

- How do we meet future water needs?

This activity has also been used with first year students framing problems that inhibit inclusive community in a dorm; students and community members framing problems in local food systems; and interdisciplinary students framing problems related to a recent flood. Across these contexts, students and community partners have learned to identify how problems are social constructions, that different people understand problems in different ways, and that it takes focused reflection and dialogue to identify and productively combine different perspectives on problems.

Activity 3: mirror, mirror

Environmental communication and problem-solving often require the contribution of multiple experts. For example, addressing issues such as unhealthy air quality requires subject-area expertise in ecology, atmospheric chemistry, urban planning, and public health administration. Designing an EC or public awareness campaign for such issues requires precise understanding and translation of high-level scientific information. Thus, projects of this nature, whether focused locally, regionally, or internationally, require the collaboration and engagement of an interdisciplinary team.

Whether with small student teams or larger research teams, it is useful to reflect upon the communication and collaboration process itself. "Mirror, Mirror" is an adaptation of Roux et al.'s (2010) reflective adaptation process and works best as an instructor- or team leader-facilitated session halfway through a project cycle and again near the end of the project. The group session begins with each individual responding to an open question about the project (e.g., "what aspects of the project have been challenging, rewarding, or most important for you in your role as a student/scientist/individual/or related role"). There are no restrictions for responses, which can range from a stream-of-consciousness-style essay to bullet points to a formal learning assessment document. Once input is collected, the facilitator reads and open codes the responses for key patterns and themes to identify concerns and celebratory moments shared by multiple team members. The codes are synthesized, and a few examples are fleshed out for sharing with the team. All responses are kept anonymous. At the reflective session that follows, the facilitator introduces the principles of "compassion" and "candor" to guide the group discussion. Then the facilitator plays the role of the "mirror" and shares a few key themes identified in the reflective essays, which become the foundation for a compassionate and candid, full-group conversation about what the mirror sees and what is missing from the reflection. "Mirror, Mirror" can help improve team communication and collaboration as the project continues. At the end of a project or semester, it can help participants and students process what they have learned and bring closure to a complex, collaborative effort.

We facilitated this activity with an international team investigating the ecological and social impacts of climate change on livestock herding communities in Mongolia. The interdisciplinary team began working on the project in 2007. After receiving a $1.5 million grant from the National Science Foundation, the team expanded and recruited more subject-area experts to participate in the data collection, analysis, and synthesis. Prior to a team "data-sharing" retreat in 2011, we circulated the list of "mirror, mirror" questions to fifteen of the core team members. Twelve members sent in-depth responses, which were then anonymized, coded, and presented back to the team members at the retreat. The main themes included relationships, commitment, inclusivity, roles and responsibilities, and overall project management. A robust conversation ensued, in which team members discussed their concerns about and challenges of conducting this type of collaborative environmental and international work as a multidisciplinary team. In a post-meeting evaluation form, one team member wrote: "You said what I was feeling. The thing that struck me as most useful [about this exercise] was [bringing] everyone together and the candidness of the discussion." After that exercise, the team's leadership included more open discussions about the teamwork process and offered regular "check-ins" at monthly meetings to reflect on team dynamics and their influence in the environmental problem-solving process.

Activity 4: shared systems mapping

This activity, adapted from Daniels and Walker's (2001) situation mapping, helps collaborators identify systems ontologies through graphic sketching and allows collaborators to expose differences in their assumptions about knowledge claims related to systems. In a recent synthesis of their twenty years of experience encouraging systems thinking processes in natural resource policy decision-making contexts, Daniels and Walker (2012, 105) note, "there may be no better way to promote breakthrough creativity than to get diverse stakeholders interacting with one another in a shared systems understanding of the situation." Like the other activities, shared systems mapping can occur at any stage in a collaboration. However, we have found that using this activity after collaborators have developed skills in identifying and negotiating differences in problems can enhance the breakthrough creativity from this particular activity (Daniels and Walker 2012).

Shared systems mapping is a process of graphically representing an issue to create a shared and systemic understanding of it. There are multiple ways to organize this type of activity, and the approach depends in part on the size of the group and the available resources. This process can be done using chalkboards, large sheets of paper, or basic computer design programs. The activity facilitates dialogue and exploratory thinking among participants. Facilitators explain that the goal is not the actual systems map, but the discussion, careful thought, and learning that accompany the mapping process. Mapping can happen early in a project or class to understand how various components, concepts, and collaborators are connected, but should continually be revisited to identify new contributions and shared learning along the way. Also, mapping can be done at a range of levels—from very simple sketches to descriptive details reinforced with available data sets and examples. The details shared and level of complexity are a direct result of the mapping team's perspective and available knowledge sets. Ultimately, the process of sharing this information and building the map collaboratively stimulates capacity for effective communication and problem-solving.

One way to extend this activity and to reinforce learning about differences in problem framing is to do Systems Sketching before Shared Systems Mapping. In Systems Sketching, people individually draw their visualization of a system. In our classrooms, we have continually been surprised at diverse ways in which people view systems. We see drawings of systems that resemble classic ecosystems, usually with a sun, trees, and arrows connecting cycles of matter and energy; the dynamic systems that Meadows (Meadows and Wright 2008) described and visualized with boxes showing inputs and outputs connected in one way and cyclical flows; and complex systems that show the influence of chaos thinking in their fractal branching and swirling patterns (Ramage and Shipp 2008). Together, Systems Mapping and Sketching helps people grapple with the complexity, controversy, and uncertainty in these messy problem contexts in ways that can promote transformative learning (Daniels and Walker 2012).

Activity 5: collaborating on a case

To address contemporary environmental challenges, citizens and scientists alike must be able to handle complexity, ambiguity, and urgent problems that have no obvious optimal solutions (Sprain and Timpson 2012). Case-based approaches use representations of problematic situations in real life and apply them as a basis for inquiry, practice, and play. Working with cases equips students and collaborators to encounter complexity, manage uncertainty, and generate innovative strategies as they develop interpersonal competence to work together in diverse teams. Collaborating on a case can be an activity that comes after students and collaborators have developed skills in the first four activities, but it can also be used as an overarching organizational strategy. As an organizational strategy, collaborators focus on working on a case together and at distinct stages use the first four activities to advance their work. As a synthesis, this activity helps people learn more about and apply framing, critical thinking, systems thinking, and deliberation.

Students or collaborators divide into groups of three or four to identify, build, and present a case study that analyzes and makes recommendations about some of the rich and difficult facets of wicked environmental problems. In a classroom setting, each group can assign several related readings prior to class and then lead classmates through a 45-minute active discussion of their case study, including a role-playing activity, and submit a digital portfolio of case study materials. Students are encouraged to identify case studies that resonate with their research, teaching, and activism interests. Case studies are assessed based on the quality of the materials collected, the relevance to course themes, the critical thinking demonstrated by students through their in-class participation and assignments, the usefulness of the solutions presented, and the polish of the in-class presentation and digital portfolio. In team contexts, team members can work to build a case directly related to their shared project, incorporating role playing as a chance to step into colleagues' shoes to consider other viewpoints and approaches to the issue at stake.

We have employed this activity, for instance, in a course on public engagement with science for graduate students in biological and environmental sciences where students worked in small, interdisciplinary teams to identify, compile, synthesize, and present a problem-focused case. Examples included issues such as international comparisons of wind farm siting processes, co-management of the Galapagos Marine Reserve, the announced closing of a long-term and important freshwater research reserve in Ontario, and Maine's Penobscot River Restoration Project, which resulted in the removal of two major dams and improved fish passage on two more. After presenting their cases, groups led the class in related interactive exercises. For example, one group set up a role play to explore decision-making about siting a wind farm. In another they developed an exercise where students received varying degrees of information about a scientific topic and had to find ways to adapt what they knew (or didn't know) to communicate with their audience, in this case a mock funding body made up of other students. Students had to work with the incomplete information with which they were provided and connect with the funding body using audience-appropriate terminology and framing and, as a result, these future scientists learned how to find persuasive strategies in the midst of uncertainty and relate the relevance of their mock research for a public audience. This activity allowed students to develop adaptive capacities that focus on accepting complexity, ambiguity, and change as they work through role plays and negotiations in real time, to consider problem framing and its impact on decision outcomes, and to strengthen their interpersonal communication skills, among other capacities.

Conclusion

This set of activities takes seriously EC's call that we develop research and praxis to understand, address, and potentially solve complex social and environmental problems (Cox 2007, Schwarze 2007, Endres et al. 2008). The activities have been developed and refined in diverse classroom and community settings (McGreavy et al. 2016), and, because of this, we are confident they can easily be adopted and adapted for use in international contexts and with diverse groups. It is through the application in worldwide settings where there is a pressing need to find ways to come together and meaningfully deliberate about issues related to climate change, social and environmental justice, unsustainable land and water resource management, and many more urgent situations, that these activities will realize their full potential as transformative praxis.

References

Bakhtin, M. (1990) "Marxism and the philosophy of language," in Bizzell, P. and Herzberg, B., eds., *The rhetorical tradition: readings from classical times to the present*, Bedford/St. Martin's Press, New York, pp. 928–44.

Cox, J. R. (2007) "Nature's 'crisis' disciplines: does environmental communication have an ethical duty?" *Environmental Communication*, 1: 5–20.

Daniels, S. E. and Walker, G. B. (2001) *Working through environmental conflict: The collaborative learning approach*, Praeger Publishers, Westport.

Daniels, S. E. and Walker, G. B. (2012) "Lessons from the trenches: Twenty years of using systems thinking in natural resource conflict situations," *Systems Research and Behavioral Science*, 29: 104–15.

Dewulf, A., François, G., Pahl-Wostl, C. and Taillieu, T. (2007) "A framing approach to cross-disciplinary research collaboration: experiences from a large-scale research project on adaptive water management," *Ecology and Society*, 12: 1–24.

Endres, D., Sprain, L. and Peterson, T. R. (2008) "The imperative of praxis-based environmental research: suggestions from the Step It Up 2007 national research project," *Environmental Communication*, 2: 237–45.

Kreuter, M. W., Rosa, C. D., Howze, E. H. and Baldwin, G. T. (2004) "Understanding wicked problems: A key to advancing environmental health promotion," *Health Education & Behavior: The Official Publication of the Society for Public Health Education*, 31: 441–54.

Lindenfeld, L. A., Hall, D. M., McGreavy, B., Silka, L. and Hart, D. (2012) "Creating a place for environmental communication research in sustainability science," *Environmental Communication*, 6(1): 23–43.

McDowell, S. (n.d) "The 'Yes, and …' approach: Less ego, more openness, more possibility," 99U. Available at: http://99u.com/articles/7183/the-yes-and-approach-less-ego-more-openness-more-possibility (accessed May 15, 2016).

McGreavy, B., Hutchins, K., Smith, H., Lindenfeld, L. and Silka, L. (2013) "Addressing the complexities of boundary work in sustainability science through communication," *Sustainability*, 5: 4195–221.

McGreavy, B., Druschke, C. G., Sprain, L., Thompson, J. and Lindenfeld, L. (2016) "Environmental communication pedagogy for sustainability: Developing core capacities to engage complex problems," *Applied Environmental Education and Communication*, 15(3): 261–74.

Meadows, D. H. and Wright, D. (2008) *Thinking in systems: A primer*, Chelsea Green Publishing, White River Junction.

Ramage, M. and Shipp, K. (2009) *Systems thinkers*, Springer, London.

Roux, D. J., Stirzaker, R. J., Breen, C. M., Lefroy, E. C. and Cresswell, H. P. (2010) "Framework for participative reflection on the accomplishment of transdisciplinary research programs," *Environmental Science & Policy*, 13: 733–41.

Schwarze, S. (2007) "Environmental communication as a discipline *of* crisis," *Environmental Communication: A Journal of Nature and Culture*, 1: 87–98.

Sprain, L. (2013) "Framing science for public action," in Goodwin, J., Dahlstrom, M. F. and Priest, S., eds., *Ethical issues in science communication: A theory-based approach*, CreateSpace, Charleston, pp. 279–86.

Sprain, L. and Timpson, W. (2012) "Pedagogy for sustainability science: Case-based approaches for interdisciplinary instruction," *Environmental Communication*, 6: 532–50.

21 Image(ination) and motivation

Mapping environmental positions in visual media

Mary Stroud

The importance of visual literacy in our multi-media saturated world is a central topic in rhetoric and writing studies, and an environmental focus in this area is more essential than ever. Indeed, students receive a barrage of visual environmental messages on a daily basis, from the polar bear perched on melting ice surrounded by a vast sea, to green labels on organic products, to the sublime landscapes of national parks captured in digital media, making the critical analysis and understanding of these visual texts of paramount importance. This exercise aims to combine prevailing scholarship in environmental communication with analytical methods espoused in social semiotic studies and offers students tools and techniques to identify the ideological orientation behind the visual messages they encounter. Through the techniques presented here, they should gain greater insight into the overall efficacy of visual media to convey environmental exigencies and be able to interrogate the ethical potential of such messages.

This assignment relies on two primary elements, a visual analysis "toolkit" derived from and a continuum map of environmental ideologies. By pairing these elements together, students gain vocabulary and skills to effectively analyze visual media. I use this exercise in a writing and rhetoric course dedicated to visual literacy. The exercise leads up to the first essay of the course in which students are asked to develop an argument by rhetorically analyzing an environmental image of their choice.

Student handout

Before beginning this exercise, please take a moment to study your chosen image. Then, fill in the blank to the following statement:

I believe the message of this image is ______________________.

Now that you have determined what you think the message of the image is, use the following toolkit and continuum to expand on your analysis. By the end of this exercise, you should be able to articulate how the image achieves its specific message, and you may find that the message you determined above changes. Allow at least 30 minutes to complete this exercise.

Visual analysis toolkit

Kress and van Leeuwen's (2006) methods of social semiotic analysis provide us with essential tools for analyzing images. Although it is necessary to simplify their theories and exercises for the purpose of quick application here (please see their text *Reading Images: The Grammar of Visual Design* for a greater in-depth study of their fascinating and seminal work), the following steps will provide you with the initial tools to understand the message of an image in deeper ways.

Step 1: Narrative Analysis through "Vectors"

When you encounter an image, your first instinct is likely to "tell the story" of what is taking place before your eyes. This is a rather challenging task when working with static visual material because it is not linear like a text or dynamic like a film, thus it is up to you to determine the significance of any action taking place. For this step, you will essentially rewrite the idiom "a picture is worth a thousand words" by using vectors to carefully select exactly which words describe the action in the frame. Kress and van Leeuwen (2006) explain that a "vector" is any line of movement in the image. By tracing out the vectors in the image and applying an action verb to each vector, we are able to "tell a better story" for rhetorically analyzing the image:

1. Scrutinize the image carefully for any vectors that might be present, remembering that vectors can be as subtle as the direction of a gaze or as overt as arms outstretched. Draw a rough outline of the main characters in the image and represent the vectors with arrows moving from the main subject of the action, the "actor" toward a specific "goal." Make special note if the goal does not exist inside the image.
2. Number the vectors, list the numbers, and then compose a sentence that captures narratively what is happening with each vector using a specific action verb for each vector. If the goal does not exist in the image, use an intransitive verb.

Once you complete this part of the analysis, move on to Step 2.

Step 2: Symbolic Analysis

Your next instinct when studying an image is likely to think about how symbols are operating in the image. For example, an eagle obviously has more meaning than simply a large bird of prey when it is coupled with an American flag, in which case its meaning expands to include virtue and patriotism. Kress and van Leeuwen (2006) explain that symbols always have a primary figure that carries the meaning and attributes that convey the meaning. Analyzing these elements discretely helps to reveal how symbols function within an image:

1 Search the image for any symbols, remembering that symbols are visual signs for abstract ideas. Write them down and then identify the carrier (the primary figure) and the symbolic attributes (the elements that convey meaning) for each symbol.
2 Given your listed symbols in Step 2.1, write out a brief analysis of the meaning of the symbols in the context of the whole image. Are certain cultural meanings afforded more or less importance? Does tension exist between any of the symbols?

Once you complete this part of the analysis, move on to Step 3.

Step 3: Modalities

Kress and van Leeuwen (2006) explain that we rely on certain visual "modalities" to recognize the reliability of messages within an image. If you have ever edited an image for your Instagram feed, you've played with "modalities," which include color saturation, differentiation, and modulation. Basically, do you make your image sepia toned or black and white? Do you highly saturate your image? Do you choose to render an image in cartoon-like primary colors? The reliability of an image, its relationship to the "truth," is dependent upon contextual expectations we have about how modalities should function. For example, we take highly saturated images as reliable on national park brochures and websites, but would balk at such photo manipulation in a news story. In the first case, we expect to be awed by promotional material for a national park but want to know the cold hard facts about a news story that takes place there:

1 Study the image and write down observations about modalities. Are the images photorealistic? Are they highly saturated? Are they black-and-white or sepia toned? Do they contain only primary colors?
2 Use your instincts here. Given your findings, what do you think you are to take as "truth" based upon the modalities of the image? What are the different contexts in which the given modalities would be perceived as reliable?

Now that you have completed these three steps, move on to map your findings on the continuum below.

Mapping your findings

The continuum below (Figure 21.1) has been derived and adapted from a variety of sources to help you think about where to locate your toolkit findings in a range of ideological orientations about the environment.

1 Map your findings along this ideological continuum. Look to your original statement about the image that you made at the beginning of the exercise.

Nature as object	Nature as resource	Nature as subject
• Humans are depicted to be "acting on" their environment. • Symbol analysis reveals nature viewed as something that can be possessed or studied. • "Truth" revealed through modalities is oriented toward human experience	• Humans are depicted to be using elements from their environment. • Symbol analysis reveals nature viewed as something that can be altered, possessed, or used. • "Truth" revealed through modalities is oriented toward human use.	• Elements from the environment are shown to impact human participants or function as actors. • Symbol analysis reveals nature viewed as something with agency influencing human subjects. • "Truth" revealed through modalities is oriented toward interactions between humans and nature, where nature has agency.
Anthropocentric		*Ecocentric*

Figure 21.1 Continuum of visually rendered environmental ideological orientations.
Source: adapted from Killingworth and Palmer 1992, and Corbett 2006.

Place a distinctive mark to denote where that message might exist in the top part of the figure somewhere between "Nature as Object" and "Nature as Subject." Then, returning to the toolkit, circle the descriptions that accurately describe your findings from each area of the toolkit (narrative, symbols, and modalities). If you feel that the categories are too restrictive, feel free to add additional descriptions or categories to the continuum. Revise it and adapt it to reflect what you see in the image.

2 Now, revisit the statement from the beginning of the exercise. How would you revise this statement based on your findings here? Rewrite it by filling in the blank to the same statement:

I believe the message of this image is ____________________.

When you have completed this exercise, be prepared to share your findings with the class.

Evaluation

The objective of this exercise is to assist students in developing critical visual literacy so that they can make convincing and articulate arguments about the images they analyze. It is also to help them understand how environmental ideologies manifest in the visual media they encounter every day and assess the efficacy of the conveyed messages. Because this exercise leads up to the first graded essay of the class, the exercise itself is ungraded and is meant to be produced during class time followed by a thought-provoking discussion of their findings. In particular, I ask students the following questions to help them extend their findings from this exercise into the creation of arguments for their essay:

- Were you surprised by what the visual analysis revealed? If so, how did you see the message of the image differently after you applied the analytical steps?
- Did the mapping prove challenging? Did the image reveal conflicting messages in terms of your initial impression of the message and what you discovered through the toolkit? Were you inspired to add new categories or revise the continuum in any way? If so, how?
- Can you identify particular stakeholders based on the message you came up with at the end of the exercise? If so, who are they? Are you one of them? Do you yourself feel moved or influenced by the message conveyed?
- If you could redesign the image for a more efficacious and ethical message about the environment, how would you do it?

The greatest measure of evaluation and success for this work would be that students gain a stronger critical awareness of the role of environmental images and likewise apply that awareness to their own activist work and attempts to visually communicate environmental exigencies most effectively.

References

Corbett, J. (2006) *Communicating nature: How we create and understand environmental Messages*, Island Press, Washington, DC.

Killingsworth, M. J. and Palmer, J. S. (1992) *Ecospeak: Rhetoric and environmental politics in America*, Southern Illinois University Press, Carbondale.

Kress, G. and van Leeuwen, T. (2006) *Reading images: The grammar of visual design*, Routledge, New York

22 Using infographics to put environmental communication into practice

Antonio López

The goal of this activity is to create a visual form of environmental communication using free online infographic tools, prompting students to put into practice a variety of concepts learned in an introductory environmental communication course. Infographics are increasingly used to explain and map environmental issues in publications (journals, magazines, newspapers) and on social networks (Pinterest, Facebook, Twitter, Tumblr, Flickr, Stumbleupon, Digg, Reddit, etc.). Learners with no experience creating infographics can successfully make one.

The planning and creation of an infographic involves visual expression and creativity, embedding basic communications skills that are transferable to other types of media (such as journalism, graphic design, or filmmaking); learners develop multifaceted communication skills as they visualize complex concepts and distill them into a simple message for a targeted audience. Students must conceptualize and make a claim about an environmental problem to demonstrate symbolic action (Cox 2009); they develop a discursive strategy based on concepts and principles of environmental communication. In particular, students combine visual and written techniques to communicate an issue and carry out research using skills such as information literacy (the ability to ascertain the validity of research and seek out sources of information) and visual literacy (the ability to meaningfully connect imagery with the communication of ideas).

An infographic can be broadly described as a form of visualized data, information, knowledge, or concepts. There are a variety of types, including maps, conceptual maps, scientific illustrations, statistical infographics (graphs, charts, infointeractives), non-statistical infographics (fact boxes, tables, non-data maps, diagrams, time), flow charts, word/tag clouds, and calendars (Lester 2014; see esp. Chapter 9 for a good overview of different types of infographics and their history). For examples to demonstrate in class, you can do a web search for infographics using the above categories and create a PowerPoint presentation for the class. When possible, combine your search terms with environment, sustainability, or related concepts to find as many relevant examples as possible.

For the assignment, students should define a particular problem or issue within the context of environmental ideologies as defined by Corbett (2006). As they research their topic, they will narrow the focus to a specific take-home message designed for the actual audience they intend to reach. They then develop a communication strategy to persuade, inform, or transform their audience. Finally, they decide on the best channels for reaching this audience and map out a strategic communication plan (which can be included in a final reflection paper).

The assignment

This assignment has the following objectives:

- Visually communicate an environmental issue that incorporates environmental ideology and discourses.
- Develop an environmental communication strategy with clear goals.
- Develop a creative and esthetic approach to communicate an environmental topic.
- Combine words, images, graphics, and numbers to illustrate key concepts.
- Strategically and purposefully use fonts and color.
- Use icons, graphics, and images based on environmental tropes.
- Incorporate clearly written elements.

Background

To perform this assignment, students should have already explored examples of environmental ideologies and discourses communicated in diverse forms of media, such as advertisements, films, popular culture, and the news. They should be familiar with the types of discourses and tropes used to communicate different environmental ideologies and claims-making positions (Hansen 2009).

Planning

For this activity, students should work through the following steps:

- *What do you want to communicate?* It is important to have a clear take-home message that can be summarized in one sentence. The best infographics offer something surprising or unexpected. Choose a topic so that viewers learn something they didn't already know. It should do more than describe *what* the problem is, and include *how* or *why* it exists and what to do about it.
- *With whom do you want to communicate?* The targeted audience determines the style of communication and the appropriate channels of communication. For example, different communication approaches will be required to promote solar power to real estate agents than to university students. Students should target an actual audience. You can either require the infographic to be published and distributed, or you can have students develop an outreach plan.

- *How do you want to communicate?* Here students use specific environmental communication strategies by selecting types of discourses and symbolism appropriate for the target audience.
- *Where do you want to communicate?* The placement of the final project should be mapped out by considering what venues are available (discussion groups, social network sites, etc.). Students can prepare press releases for their infographic that are targeted to different kinds of media organizations.

Creating the infographic

Fortunately, there are a number of free online services and tutorials that any novice can use to create an infographic (follow this link for assignment resources and slideshow with sample infographics: http://ecomedialit.com/infographics/). After designing a communication strategy, the guidelines for the infographic are:

1 Create a title and subhead that encapsulates the main idea.
2 Include at least eight facts/statistics and use at least five sources for research.
3 Include at least three distinct sections (distinguishable by concept and separated visually with color, lines, or graphics).
4 Combine words, images, graphics, numbers, fonts, color, icons, and visual metaphors.
5 Cite all sources at the bottom.
6 Distribute the final project online and monitor its reach (optional).

Students should be encouraged to sketch and draw their concepts before using online tools.

Assessment should be based on whether or not all the required elements from the instructions were included and how successfully they implemented their environmental communication goals. I have students write a three- to five-page reflection paper that describes their process and goals, including their main take-home message, target audience, where it will be published and how it will be publicized, and environmental communication strategy (tropes, discourses, ideologies). I prompt them to write a narrative description of every step they take and their decision-making process, discussing what was difficult or challenging, what worked, and whether there was anything they wanted to do that they couldn't do because of technical reasons. This enables me to understand what their intentions were and how closely the final product matched their desired impact. Rather than assess on technical ability, I look for quality of information and research, clarity, simplicity, esthetic choices, creativity, effort, and concept. Regardless of background, every student that I have assigned this to (more than 100) has succeeded in creating an interesting project. Many find it challenging, but most respond positively to the experience and often comment that they appreciate the opportunity to put theory into practice, while also learning a practical skill they can apply in other classes or in a work environment.

References

Corbett, J. B. (2006) *Communicating nature: How we create and understand environmental messages*, Island Press, Washington, DC.

Cox, R. (2009) *Environmental communication and the public sphere* (2nd edn.), Sage Publications, Thousand Oaks.

Hansen, A. (2009) *Environment, media and communication*, Routledge, New York.

Lester, P. M. (2014) *Visual communication: Images with messages*, Wadsworth/Cengage Learning, Boston.

23 Analyzing news media's incorporation of animal and eco perspectives

Carrie P. Freeman

This out-of-class assignment helps students discover for themselves the extent to which environmental and animal protection issues are addressed in the mainstream news media, as a way to explore media agenda setting and news construction and framing. It may result in a critique of agenda setting priorities in cases where important environmental issues are being ignored or marginalized by trivial topics or solely anthropocentric concerns. Students' findings will vary based on the news cycle at the time. In cases where an environmental story dominates the news cycle (e.g., another oil spill), instructors could discuss the need for variety/diversity in stories across a wide spectrum of issues (or the need for more thematic rather than episodic coverage). The assignment also serves as an opportunity to debate what counts as "environmental" as well as how environmental and animal protection angles could factor into many existing news stories where they are currently left out. This helps to show that the interests of nature and nonhuman species (along with the interests of humans who are disproportionately affected by pollution or climate disruption) should be brought more frequently into mainstream discussions on related issues that affect these entities. Guidelines in www.animalsandmedia.org particularly encourage journalists to expand their consideration of who counts as a source and whose interests and perspectives should be included in news stories (i.e., nonhuman animals). This assignment gives students the opportunity to put the animal and media style guidelines into practice by making recommendations on how to improve existing news stories in terms of respectful inclusion of nonhuman animal interests.

During the weeks that environmental communication teachers are addressing the news media, this could be a homework assignment to help put the course readings into action. Alternatively, this could be used as a make-up homework assignment to replace a zero, or extra credit. If instructors want to infuse some discussion of current news issues into every class, they could assign this as extra credit to a different student each week, having them devote five to ten minutes discussing their findings to the class (especially for the agenda-setting portion). To prepare, in addition to reading environmental communication textbook chapters on journalism, agenda-setting, and framing, students could be assigned Freeman et al. (2011). Students who need more grounding in communication theories of social construction and discourse should read the first chapter of Hall (1997).

Assignment instructions for students

Part 1: News media topic selection (agenda-setting)

Go to any major/mainstream news organization's website and make a cursory review of its range of news stories and topics that week across all its sections/beats (international, regional, politics, business, sports, science, technology, living, dining, editorial, etc.). Address the following questions:

1 In terms of agenda-setting, what are the main topics that seem to be given the highest priority and most attention this week?
2 What stories (if any) did you see that you think qualify as being related to the environment and in what sections were they? Estimate what proportion of overall news stories seemed to involve environmental topics.
3 What stories (if any) did you see that were related to nonhuman animal issues and in what sections were they? Estimate what proportion of overall news stories seemed to involve animal topics.
4 Given your findings to the first three questions, what comments or concerns do you have about how the media is setting the agenda?
5 Identify several missed opportunities for where "the environment" (or perspectives related to nature, nonhuman animals, or environmental justice for humans) could have been a relevant component in a current news story/topic, but was not. What is your argument for how it would have fit and been newsworthy?

Part 2: News story construction and framing

Find a current news story that relates to nonhuman animals.

1 Using the media style guidelines at www.animalsandmedia.org/main/journalism-guidelines/ analyze the news story in terms of how it followed or didn't follow the guidelines. (Give the analysis a few paragraphs.)
2 What specific improvements in social construction and framing of issues would you recommend to the news organization based on these guidelines? Do you have any additional recommendations based on your own value system?

Learning objectives and evaluation

For Part 1, the first three questions are descriptive in nature, with the objective that students assess the news landscape from an agenda-setting lens to gage the extent to which the news cycle prioritizes issues affecting nature, ecosystems, and other animals. This also allows students to define what "environmental" means and encompasses; they may use a definition from a textbook or perhaps come up with their own definition. This is followed by a critical thinking question

(question 4) where students can share their own opinions and concerns about their findings from questions 1–3. During this analysis, question 4 gives them the opportunity to draw upon assigned readings on journalism (most likely from an environmental communication textbook). From this, the teacher can assess the extent to which the student engaged with and comprehended assigned readings. Question 5 then asks students how they could bring environmental and animal protection concerns and perspectives into the existing news stories. The objective here is to encourage creative thinking and problem-solving by putting them in the position of a journalist in constructing a news story.

This broad survey of the news landscape ties into Part 2, where the focus narrows and they are asked to closely analyze the construction and framing of a specific news story (this one about nonhuman animals) and improve upon it. The objective is to demonstrate that the news is not simply an objective record of events, but that it is a chosen/selected construction of a narrative based on viewpoints, values, and social norms. In these last two questions, the style guidelines at animalsandmedia.org are used as the criteria by which students assess how respectful and inclusive a news story is. The last question, asking them to reconstruct the news story, is broad enough that they can disagree with parts of the animalsandmedia.org guidelines, as it also asks them to consider their own value system in the critical analysis and narrative choices.

The teacher should evaluate students' success based on the extent to which they answered the questions thoroughly and applied critical thinking, creative problem-solving, values, guidelines, and concepts from the readings, where they apply. Students should demonstrate a concern for and offer ideas about how journalism could better incorporate nonhuman animal and environmental interests and issues into the news media to challenge and alter the anthropocentric status quo.

References

Freeman, C. P., Bekoff, M. and Bexell, S. (2011) "Giving voice to the 'voiceless': Incorporating nonhuman animal perspectives as journalistic sources," *Journalism Studies*, 12(5): 590–607.

Hall, S. (1997) *Representation: Cultural representations and signifying practice*, Sage Publications, London.

24 Newschart assignment

Karey Harrison

In this Newschart assignment[1] students collect news articles, analyze them, and record the details. Part of a larger assignment developed as an alternative to assessing students with an exam and an integral part of my overall course design it helps students develop research skills that enrich their understanding and grasp of environmental communication. The Newschart is dramatically successful in both raising students' general awareness and understanding of public, political, and economic affairs. It also increases their understanding of theoretical and conceptual ideas introduced in class. Consequently, students develop more sophisticated analyses of media reporting in relation to environmental issues. Course materials raise issues associated with "humans as part of" or "outside of nature," with globalization, and with inequalities of class, race, and gender. Students discuss and reflect on these issues in their assignments. Finally, students are encouraged to relate course readings on these topics to the newspaper articles they analyze in their Newscharts.

Newschart

The Newschart is as much a formative learning activity as it is an assessment task. The process of selecting, reading, categorizing, and analyzing articles they find ensures students reflect on the concepts and ideas introduced in the course and leads them to revise their preconceptions. By determining what counts as an appropriate article, finding articles, evaluating articles, allocating them to columns 3 and 4 in the Newschart, analyzing and identifying patterns in reporting, and justifying and commenting on their interpretation, students develop and demonstrate their research skill capabilities (Willison and O'Regan 2007). Students download, duplicate, and enter their analysis into a template as shown in Table 24.1.

The criteria for selecting news articles is intended to be at odds with the news media tendency to treat humans and the economy as separate from the environment and nature. Petrol prices, road works, or shopping and housing developments are rarely labeled as "environmental" but, nevertheless, they have environmental implications. An ecological perspective presupposes an understanding of interdependencies and connections so all human activities are seen

Table 24.1 Extract from 2010 Newschart entry for Week 5

Column 1	*Column 2*	*Column 3*				*Column 4*			*Column 5*
Week Number 5	List main news value article conforms to	**View of Environment:** Check which columns below apply							Explain/justify the cell you checked in Columns 3 & 4
		Economics given priority				Ecosystem perspective			
		Pro-growth	Ignores environmental costs of development	Promotes development, consumption	Nature as enemy	Shows costs of growth	Promotes "eco" product	Considers non growth, reduced consumption	
Local paper: *Redcliffe & Bayside Herald* Wednesday, August 18, 2010									
Council water war	Proximity		X						Focus on costs to council and ratepayers.
Energy scheme sets milestone								X	To achieve Q2 target.
State paper: *The Courier Mail* Monday, August 16, 2010									
Deluge yet to peak has left 20 million homeless	Timeliness				X				Worst floods in Pakistan, takes lives, causes spread of disease and devastates livelihoods.
National paper: *The Australian* Monday, August 16, 2010									
Pakistan floods: 20m homeless	Timeliness				X				Worst floods in Pakistan, causes spread of disease and homelessness.

Comments for week: As the week leading up to the election, many articles focused on this. Of interest is the increasing difficulty in deciding what constitutes an article about the environment. Investigative journalism into articles about refugees and boat people, for example, might uncover an environmental element that is not obvious in most articles. The articles in the business sections of the newspapers could also largely be included as they focus on growth and increased consumption.

as having implications for the biosphere. The media, in contrast, categorizes most human activities as occurring outside the environment, and news reports on environmental issues generally represent them as external to social, political and economic institutions.

Articles classified by news media as "environmental" tend to reinforce the perspective of humans separate from nature. Students are asked to collect each week ALL the news stories on issues with an environmental impact from a local, a state, and a national newspaper. Students are asked to distinguish between environmental impact and articles labeled by the media as "environmental" stories. This encourages them to reflect on humans as part of the environment as compared with the separation of environment and society in media reporting. To decide what is an environmental issue students are given specific reference to the view of humans as part of a complex ecosystem discussed by Lovelock (1988).

Learning activities

In supporting course material, students are introduced to the anti-environmental biases of news values, as shown in Table 24.2. Environmental problems such as human-induced climate change, symptomatic of long developing processes, rather than the consequence of a one-off event, do not lend themselves to reporting of "facts" or "events." It is difficult to find a news peg for something that is not an event. Processes, by their nature, are neither timely (they have been going on for some time) nor unusual—something that has been going on for some time is "business as usual," the opposite of "unusual." To analyze social and economic structures or evaluate alternative explanatory models of complex ecosystem processes such as climate change, journalists need to go beyond the "objective facts" or competing "claims" to which "objective" or "balanced" news reports are limited (Boykoff and Boykoff 2004).

Box 24.1 News value learning activity

In class students break up into small groups to examine newspapers. Each group practices identifying stories with environmental impact, and the most important news values in each story, justifying their selection and news values in a report to the whole class. Students then discuss the relative frequency of news values in relation to what those values privilege and marginalize. They reflect on the relation between news values and bias as they record items in their newschart, and decide which of columns 3 and 4 in Table 24.1 the item belongs in. In sorting and analyzing articles students have chosen for their newschart, they discover for themselves how the mass media is biased against the view of "humans as part of nature" and do not have to take my word for it while also practicing how to conduct simple data collection and analysis for media research.

Table 24.2 News value and systemic bias of journalistic norms

News value	*Privilege*	*Marginalise*
Journalistic norms	Mechanistic causation	Complex systems (environment)
Facts	Events/quotes	Structure and processes
Conflict	Us vs. them	Analysis
Prominence	Powerful, wealthy, famous	Subordinate groups and interests
Timeliness	Short term	Future consequences
Proximity	Parochial	Big picture
Unusual	Dramatic change	Developmental increments

Learning outcomes

Column 5 in Table 24.1 asks students to justify or provide the reason each article belongs in the sub column they have selected under columns 3 or 4. These reasons allow the instructor to evaluate both their understanding of the key concepts discussed in the course and the quality of evidence they provide in support of their judgments. The sort of insights students get from completing this assessment activity are shown in the "Comments for the week" at the bottom of the one week entry from a typical sample student Newschart shown in Table 24.1.

Working through the completion of the Newschart each week generates a process of conceptual transformation. Besides providing a tool for assessing their understanding of the course, this assignment provides a record of this transformation of their understanding and thinking about the media and environment. Because students are asked to relate readings and news articles to their own experiences, they regularly report engaging with family and friends on issues raised in the course. As their own perspective shifts, they report on attempts to influence the perspectives of those with whom they associate. They also report on how the readings have led them to reflect on and change their own practices, including changes to consumption, political views, and activities.

Note

1 This Newschart is part of the assessment for a course on "Communication and Environment" at the University of Southern Queensland (USQ) that was first offered in 1992, immediately following the negotiation of the United Nations Framework Convention on Climate Change (UNFCCC) at the Earth Summit in Rio de Janeiro, and that I continue to teach in updated form today.

References

Boykoff, M. T. and Boykoff, J. M. (2004) "Balance as bias: Global warming and the US prestige press," *Global Environmental Change*, 14(2): 125–36.

Lovelock, J. (1988) "Man and Gaia," in Goldsmith, E. and Hildyard, N., eds., *Battle for the earth: Today's key environmental issues*, Child & Associates, Brookvale, pp. 51–64.

Willison, J. and O'Regan, K. (2007) "Commonly known, commonly not known, totally unknown: a framework for students becoming researchers," *Higher Education Research & Development*, 26(4): 393–409.

25 Speaking for/to/as nature

Maggie Siebert

This in-class activity arose as a way to explore the deep and often hidden meanings and associations we have with/in our environment. Cox (2007, 12) states, "Our ideas, beliefs, attitudes, policies, and practices involving the natural world and environmental problems are mediated by systems of representation—by human communication." In this activity, students explore speaking for, with, and as the more than human world. In doing so, they personally personify and communicate their ingrained associations with environment, more than human nature, and the paradoxical question "What is natural?" or "How would it feel if I were another being/entity in this world?" The activity unearths and helps question the hidden meanings, rhetorical uses, and associations we have for and within our wider environment, what is alive in it, what is not, what we disdain, and what we have come to associate with love or nostalgia.

In the activity, students are asked to represent and experience what it is like to be an assigned entity and/or process. Students place a picture, which everyone but they can see, onto their forehead. Examples of these entities can range from "cancer" to "kitten," and some suggestions are included in the activity instructions below. The activity allows participants to develop a personal connection, or feeling, of what it is like to be other than human. As participants represent and experience how it must feel to exist differently, their objective and subjective perspectives begin to dissolve and lead them toward an awareness of ways cultural and societal frames may have assigned them an anthropocentric perspective. This activity intentionally asks students to "speak" for, with, and as an other, developing an embodied perspective, just as if they were the "2,000-year-old redwood forest" or the "fracking site" they are representing. Guide participants using the instructions provided, to explore the following learning objectives:

- Develop awareness and insight into our shared anthropocentric mindsets;
- Question biases;
- Reframe personal positions on how we speak for/to/as nature.

Students may realize, and begin to play with, the underlying complexity of how humans assign value to all non and more than human objects, entities, and processes in our environment.

This activity was inspired by close readings of articles that promote the ethical responsibility of speaking up for the more than human environment, and questioning and problematizing what happens when we do "speak for" the non-human world (Freeman et al. 2011; Milstein 2008). The activity asks: When is it our ethical necessity to become a vocal agent for our environment, or when should we give "voice to the voiceless" (Freeman et al. 2011)?

Preparation

Beforehand, prepare cards (about the size of a large index card, examples included below) to each have a picture of something found in our environment and labeled with the most concise word or phrase for it. Possibilities are endless, but some examples are puppies (and other cute animals such as panda bears), recently exploded volcanoes or recent earthquakes, landfills, home, illnesses, the city you live in, genetically modified corn, plastic, factory farm raised cattle, coal, etc. If possible, pair off the cards in advance to create a tension between the two entities represented by the pair of participants. Plan to allocate about 30 minutes of in-class time for this activity.

Activity

Divide participants into pairs. Give each participant one card, face down. Instruct them not to look at their own card and to place it on their forehead. For example, one person has a giraffe on their forehead, while the other person has e. coli. The partners then ask each other questions about the entity and/or process they are representing to guess what "they are." Instruct the participants that the point and tension of the activity is to guess what you represent while concurrently speaking for the entity you represent. During the activity, participants will very likely reflect to each other their deeply held ecocultural perspectives and beliefs. For example, when other participants look at them as if they are cancer, they may embody the feeling of being despised. As students reach the point where they understand what they are representing, they often experience a profound shift in awareness of what they previously held as a belief or assumption regarding that entity and/or process. Ideally, they also experience realizations about both the interconnection of entities/processes and ecocultural (often anthropocentric) framing. This shift in awareness, and hence consciousness raising of our deeply seated perspectives, is what the activity is designed to provoke.

Suggested question prompts for paired participants include the following:

- What do I look like to you?
- Do I make a sound? If so, what does my voice sound like? Could you speak for/to/as me?
- Do I regenerate?
- Would you eat me?

- Would you like to see me in my natural habitat? In a zoo? What is my natural habitat?
- Am I valuable? Would you use me as a resource? Am I worth money?
- Would you prefer me dead or alive?
- When you think about me, how do you feel?
- If you wanted to tell me something, what would it be?
- Why do you feel the way you do about me?

Evaluation

Evaluation begins with a debrief where partners take off their cards and discover "who they are." Group discussion then can be guided by the following questions:

1. Is there inherent and equal value to all life?
2. Is everything in the more than human world alive?
3. Is all life "animal"? Is all life "natural"? If not, what is it?
4. How would you "speak" for each of these entities? What are the processes these entities use to communicate? How can you listen to these entities?
5. Is there an ethical necessity to "speak" for more than human life forms and processes, such as those on your cards or others? If so, when?
6. Do all life forms have a purpose? Is that purpose anthropocentric? Utilitarian? Ecocentric? Explain.

Sample cards

Figure 25.1 Koala bear.

Figure 25.2 Oil rig explosion.
Source: Pixabay. Free for commercial use under CC0 Public Domain.

Wrap-up

Have participants convene as a class and share their experiences. The facilitator should note if any perspectives have shifted. Share any newly discovered insights through the experience of becoming another entity/process, and how it felt to speak for/as/to it. What implications does this have for environmental communication as a field, for you as a student, or for you simply as a person, alive, in a world that is often beyond one's ability to control, modify, or fully love and experience?

References

Cox, R. (2007) "Nature's 'Crisis Disciplines': Does environmental communication have an ethical duty?" *Environmental Communication*, 1(1): 5–20.

Milstein, T. (2008) "When whales 'speak for themselves': Communication as a mediating force in wildlife tourism," *Environmental Communication*, 2: 173–92.

Freeman, C. P., Bekoff, M. and Bexell, S. M. (2011) "Giving voice to the 'voiceless': Incorporating nonhuman animal perspectives as journalistic sources," *Journalism Studies*, 12(5): 590–607.

26 Creating emotional proximity with the environment

Maria Clara Valencia

In 2008, after several years working for the main newspaper in Colombia, *El Tiempo*, I felt the need to take a break from hectic urban life. I decided to move for a while to Ilha Grande, a natural reserve island in Brazil. That experience changed my life by reviving in me the meaning of life and happiness. Also, after spending several months among monkeys and turtles, I realized I wanted to dedicate my life to the environment.

When I returned to Colombia, I focused my journalistic efforts on environmental reporting, which I continue to this day. In 2011, I was hired as a journalism professor at Universidad Tecnológica de Bolívar (UTB) in Cartagena de Indias (Colombia). At UTB, I created the first environmental journalism course as part of the undergraduate program in Social Communication.

I begin my classes by asking students to create and reflect on maps of meaningful places, such as places where students have been happy. With these maps, we talk about environmental problems that could ruin that happiness. This activity allows them to connect their own experiences with the environment and makes learning more implicative and motivating. I use my own experiences to illustrate this. For example, since I arrived in Cartagena, an otherwise beautiful touristic city, I have realized garbage has become such a regular feature of the landscape that locals no longer notice it. I focus my environmental journalism class on looking at such local problems to understand the global phenomena, such as waste disposal. What does garbage have to do with our daily lives? Could water pollution ruin any of our routines? Creating a personal map of a place meaningful to the learner provides a visual way to connect the environment with our lives and generates personal commitments. This, in turn, may lead students to creative ways of telling stories that highlight proximity and connection. Some ideas on the importance of this kind of activity and about training in environmental journalism can be found in De la Torre and Violant (2001), Larena (2009), España and Prieto (2009), and also in FNPI (2006, 2009).

Instructions

The activity starts in groups, with each student drawing her or his own map and identifying in them places where they have been happy. The group aspect of the activity provides a space for reflection and dialogue about the maps. Some students draw the map of the city and put their favorite places in it; others just draw a less literal and more mental map where they emphasize the situations and places where they have been happy. Afterward, we reflect together about what we do in those places that make us happy and how happiness could be affected by environmental impacts. This helps us to focus on problems that are local, generate intimacy with the issue, access our emotions, and take a critical look at our surroundings. For example, if students enjoy going to the beach, water pollution or sea level rise could affect that joy. If they like to ride a bike, extreme weather conditions or lack of water could ruin that joy.

Connecting learners' personal experiences to everyday "nature" and to environmental problems through creating this map is a way to create mirror effects. The effects help them, as journalists, reflect on the power of environmental communication in ways that can lead to creating stories that deepen into community matters; these are stories that relate to people like the writer, and you and me.

Starting from that personal perspective, we discuss how we can create articles or news stories about the implications of these environmental problems we see reflected in our maps. The idea is to think of stories related to our daily lives, such as sports, food, fashion, agriculture, or economy, with an environmental lens. In this way, we discuss new ways of communicating about the environment beyond the typical stories we usually read in the environment section. In the process, we start clarifying the difference between concepts such as global warming, climate change, greenhouse gases, etc., always linking the global with the local with examples that link to their daily lives.

The exercise helps students think of different angles to write their stories. For example, one student wrote about water scarcity based on her vacation experience. She went to a youth camp when Colombia was in the middle of the driest season ever. At the beginning, campers were asked to take 2-minute showers, but a few days later they were asked to take them every other day. This experience led to an article (González and Valencia 2016) about water scarcity in Colombia, its implications nationwide, and the impact of the lack of water in everyone's life.

My main strategy has been to try to instill in students emotional proximities to nature and to connect environmental issues to their daily lives. My motto for teaching is: "No one cares for what they do not love, and no one loves what they do not know." According to Vázquez and Manassero (2007, 428), "the teaching of science has been dominated by the traditional attention to the rational processes and forgetfulness of the affections" but "emotions are very important for everything that happens within a class at any age." Education works better and the learning process is more efficient when head and heart work together. Through this exercise I want students to connect theory, technique, practice, and emotions. For me, living in Ilha Grande was a life changing experience. It is my hope that my students' environmental encounters also change their lives and have the effect of changing others' lives, as well.

References

De la Torre, S. and Violant, V. (2001) "Estrategias creativas en la enseñanza universitaria," *Creatividad y Sociedad*, 3: 21–47.

España, E. and Prieto, T. (2009) "Educar para la sostenibilidad: El contexto de los problemas socio-científicos," Revista *Eureka sobre Enseñanza y Divulgación de las Ciencias*, 6(3): 345–54, Didáctica de las Ciencias Experimentales, Universidad de Málaga.

FNPI (2006) "Relatoría del taller de cobertura de temas ambientales con Arturo Larena: La Naturalea del periodismo ambiental," Fundación Nuevo Periodismo Iberoamericano. Available at: www.fnpi.org/recursos/relatorias/taller-de-cobertura-de-temas-ambientales-con-arturo-larena-2006/ (accessed October 5, 2016).

FNPI (2009) "Los retos del cambio climático: Relatoría del taller de cambio climático," Fundación Nuevo Periodismo Iberoamericano. Available at: www.fnpi.org/comunidades/los-retos-del-cambio-climatico (accessed October 5, 2016).

González, J. and Valencia, M. C. (2016) "Cuatro días de aventura y sequía en El Socorro," *Vanguardia.com*. Available at: www.vanguardia.com/santander/comunera/359451-cuatro-dias-de-aventura-y-sequia-en-el-socorro (accessed October 5, 2016).

Larena, A. (2009) *Guía para periodistas sobre cambio climático y negociación internacional*, Secretaría de Estado de Cambio Climático España, Ministerio de Medio Ambiente, y Medio Rural y Marino SEJ (ND) Climate change initiatives Society of Environmental Journalists. Available at: www.efeverde.com/wp-content/uploads/2009/11/Gu%C3%ADa_periodistas_sobre_CC_tcm_arturo-Larena.pdf (accessed October 20, 2016).

Vázquez, Á. A. and Manassero, M. A. (2007) "En defensa de las actitudes y emociones en la educación científica (II): Evidencias empíricas derivadas de la investigación," *Revista Eureka sobre Enseñanza y Divulgación de las Ciencias* 4(3): 417–41.

27 Growing up with animals (on screens)

Gabriele Hadl[1]

All media experiences teach about environment (Lopez 2014). The media technologies and contents we live with all contain messages about humans' relationship to the natural world, though rarely ones that stand up to scrutiny. Nonhuman animals do not speak to each other in English. There is not really a soothing flute playing when you walk along a forest path. You cannot hop around in outer space without resources from the Earth. Famous Japanese gardens are not exactly free of people. Few dairy cows run around on spacious family farms with red barns and silos. And whether all living things exist merely in relation to humans is at least debatable. Yet such representations and the devices that deliver them to us influence our sense of self and our perception of the world around us.

In this activity, participants list their most memorable media experiences and screen time vs. outdoor time they have logged since early childhood. They then explore in a group whether and how these experiences relate to their values regarding humans' place among other living beings and to their own identity in relation to environmental issues and the natural world.

Context and possible applications

This activity is very basic and makes a good ice-breaker. It assumes little knowledge or preparation on the part of the participants and can be conducted outdoors. It is suitable for any age group and works well with participants of different backgrounds and ages, as it has a very democratic structure. There are no right or wrong answers, and the sharing of experiences creates trust and community. It also helps the facilitator get a sense of the kinds of attitudes present in the group. In the context of participatory syllabus design, it can be used to identify areas participants want to explore further.

In Japan, where I teach, there are very few opportunities for teaching environmental communication. I developed this activity from the My Media History/Our Media History workshop designed by my late mentor, media literacy activist Midori Suzuki. It was designed to help participants understand the concept of "living in a media society," as a society in which media contents and technologies heavily structure our perceptions of ourselves and the world.

I have used it to shoe-horn environmental communication into places where it is not expected—for example, introductory sociology courses, citizen study groups, one-off workshops for high school students. For teachers and students who have some interest in environmental and/or media issues, it is a low-stakes way of getting their toes wet. It could even be used in a team-building corporate workshop or an urban youth group. In a university seminar, it can help students identify topics for research with which they have a personal connection.

This ecomedia literacy version of My Media History/Our Media History retains the fun and emotionally invested qualities of the original, but instead of "making media more interesting" (Lopez 2014, 3), it aims to lead to critical reflection on human/non-human relations. It asks participants to notice how much (or little) time they spend experiencing without media, and to analyze the myriad messages about "the environment" and living beings embedded in the media products with which they have grown up.

Planning and facilitating this activity

Participants will spend about 90 minutes on this activity. The facilitator should also budget at least 30 minutes more of preparation time. It can be done with as few as 6 or as many as 100 participants.

Objectives

Ecomedia literacy integrates principles from media literacy, ecoliteracy, and environmental communication (Lopez 2014). This activity should deepen participants' (including the facilitator's) understanding of:

- the concepts of nature time and screen time;
- how our use of media (content, type of technology, and length of time) shapes our environmental attitudes; and
- ways representations of non-human nature include ideological messages.

Skills to be developed include:

- taking a role in a group activity;
- listening, speaking, taking notes, and presenting;
- integrating new information; and
- composition skills.

Preparing

The facilitator gives a worksheet with a grid to each participant to fill out. Horizontal rows track the participant's age and education. Vertical columns start with "outdoor time" and "media time," in which participants fill in how much time they spent outdoors versus media at each age, followed by three columns asking about what kinds of media technologies were dominantly used (TV, X-box, cinema, magazine, etc.), what kinds of content the participant most remembers (names of specific TV shows, bands, movies, comics, etc.), and what kinds of role nature and non-humans played in those contents. Participants are asked to fill this out ahead of time, if possible, and bring visuals (printouts or objects) representing their most impressive experiences (e.g., their favorite cartoon characters, their most-used technological gadgets).

If the workshop is too homogenous (participants from very similar age and cultural backgrounds), the facilitator can introduce contrasting eco media histories from other people. Alternatively, participants could be asked to get people in their community to fill out these sheets and bring them in, either as a preparatory or extension activity. If, on the other hand, the group is too diverse and participants have few common experiences, they should bring many visual examples of the media contents and technologies that have played an important role in their lives.

For higher-level academic courses, readings can be assigned beforehand on environmental ethics or the nature deficit debate. For the experienced environmental communication researcher, it may be tempting to use this activity to teach about well-researched topics. However, I find I often learn from "naïve" participants about under-researched issues and rarely considered perspectives. Rather than assigning readings beforehand that steer the activity toward a predetermined goal, it may be better to look for readings afterward and/or direct the participants toward doing their own research on the topics raised.

Sharing in a group

When participants have their personal media history sheet and visuals prepared, they work in groups to make posters representing highlights from members' experiences. These may be a big, combined version of the individually filled out worksheets or something more freely arranged. Each group should have poster-making materials and a surface to work on, e.g., a table or, better yet, a workspace on the ground. This step is best conducted outdoors. In Japan, I like to use a big picnic sheet (reminiscent of a cherry blossom party) to create a democratic atmosphere and a sense of community. If participants are new to group work, the facilitator should prepare a sheet with guidelines for effective group work. Drawing lots can be a democratic way of distributing responsibilities whereas self-nominations tend to reinforce gender and other (self)stereotypes.

Reflecting in a group

In this step, each group puts the finished poster somewhere where all in the larger group can see it and see if they can discover any common patterns and new issues. A short break can be taken to walk around and look at all group posters.

With inexperienced participants, the facilitator should prepare a discussion sheet for each group with instructions and questions such as:

- Looking at the columns of outdoor time and media time, what communalities and differences were there among members?
- What do you think are the causes and effects of different member's screen time/nature time?

The key is to help participants focus on time spent outside (far left column) and the role of nature (far-right column). If the facilitator limits talk about favorite technologies and contents (the two middle columns) to a minimum (if necessary with a stop-watch) this can help forestall devolution into tech geek or fan talk.

In general, this workshop benefits from open-ended and non-didactic questions, as long as reflections stay on the topic of how media shape our experience of nature and the environment. However, if desired, the facilitator can use questions to steer the discussion towards specific issues, such as anthropocentrism, the nature deficit debate, the effects of environmentalist or activist media, etc. The same can be achieved by assigning readings beforehand and/or as a follow-up activity.

At the end of this step, each group should summarize and prepare to present the results of their discussion, e.g., by creating another poster, adding more parts to their first poster, or helping their note taker write a summary of the main points onto a piece of paper. The designated presenter should be coached by all group members.

Presenting to other groups, wrapping up and reflecting

Next, each group presents the results of their sharing and reflections. If they have a poster from the previous (reflection) step, this can be displayed. Individual participants should be encouraged to take notes of other groups' and their own group's presentations. With many groups and sufficient time, a poster-presentation format can be arranged where all groups' designated presenters are positioned around the room with the poster and other participants mill around to hear different presentations. Groups can also be asked to respond to each other.

After this, the facilitator can comment, drawing out salient issues, and encouraging participants to investigate the meanings of terms such as anthropomorphism, zoomorphism, anthropocentrism, or human supremacism. However, in a traditionally hierarchical learning environment, the facilitator should take care not to wrap up too tightly, as participants may perceive a clear-cut conclusion as "authoritative" rather than developing their own ideas in new directions. Finally, participants write a short essay or report about what they gained from the workshop.

Assessing outcomes

It can be difficult to assess exactly how much a participant has contributed to a group activity, though peer and self-evaluation can serve as proxy where direct observation is impracticable (as it is with more than ten participants). Listening and helping others express themselves, taking responsibility for group activities, and supporting others should be valued at least as highly as speaking up. Having an extrovert personality should not automatically lead to a high evaluation. The essay or report can be graded on parameters such as how well the participant integrated and developed topics raised in the group work, if they appear to have gained an understanding of "media society" (e.g., by being able to identify any connections between specific instances of media use and certain attitudes or opinions) and of "human/animal relations" (responding to questions in the third column). In terms of skills, standard measures of good academic writing include formatting, structure, terminology, and logic.

Looking for opportunities to deepen the learning

The outcomes of group sharing and presentations are often delightfully unpredictable and even the seasoned environmental communication educator can look forward to new angles on familiar issues. That said, sharing and reflection sometimes stall at a superficial level, especially with inexperienced groups. The facilitator can use different moments in the workshop to move to a deeper level.

For example, a group may find that most members' screen time has increased since childhood, or that screen time exceeds nature time at a certain age. The facilitator may encourage them to identify the factors behind this phenomenon, such as individual, social, and technological. With advanced groups, this may also be a moment to refer them to Edith Cobb's (1977) research about the importance of direct nature experience or introduce them to the arguments of the No Child Left Inside movement (Louv 2008).

Another issue participants may notice is that animals (and inanimate objects) often talk and walk in small children's media, especially animation. This can be used to explore the concept of anthropomorphism. Why is this so pervasive in children's media and what does it mean for our relationship to animals and nature? Reflecting on media for adults, if participants note that animals and nature usually serve as a mere backdrop for human activity, this can be a moment to introduce them to the debates around anthropocentrism (or, for high level participants, to direct them towards researching the issue). Sometimes participants comment on the complete absence of things natural in certain genres such as science fiction movies or SF-themed video games, which can provide an opening to discuss to what degree human activities can actually be sustained without planet Earth. Participants also sometimes share media experiences that raised their awareness of environmental issues or deepened their love of nature. Others may report having very limited media exposure through their lives and instead reference direct experience of non-human beings. These participants may be in the minority and may need encouragement to share their experiences. Make sure such perspectives are valued and affirmed.

This activity helps participants take back their own experiences from the cultural industries. The sharing and reflections may empower them to identify messages about ways "nature" is embedded in their favorite media and to reflect on how these messages fit (or conflict) with their own attitudes about things natural and about environmental protection.

Note

1 This research was supported by a regular faculty grant of Kwansei Gakuin University and a research grant from the Midory Suzuki Media Literacy Fund.

References

Cobb, E. (1977) *The ecology of imagination in childhood*, Columbia University Press, New York.

Lopez, A. (2014) *Greening media education*, Peter Lang, Zurich.

Louv, R. (2008) *Last child in the woods: Saving our children from nature deficit-disorder*, Algonquin Books, Chapel Hill.

28 The student-run collaborative environmental communication blog

Katherine M. Cruger

In our convergence culture, it is common for students to create blogs as part of course work in writing or communication classes. However, while many have lauded social media as a liberatory and democratic outlet, individually written student blogs are often no more collaborative or interactive than academic papers (Gladwell 2010; Golding 1998). Students abandon the unread blogs immediately after the assignment is finished and seldom engage one another's work.

Courses in environmental communication demand that students consider communication as both *pragmatic* and *constitutive*, meaning environmental communication serves instrumental and practical functions such as constructing an informed citizenry, persuading decision makers, and solving concrete problems (Cox and Pezzullo 2016). When students engage with real (albeit digital) rhetorical audiences to influence perceptions on environmental issues, it brings this definition of environmental communication to life (Bitzer 1968; Cox and Pezzullo 2016).

This project works best as a semester-long undertaking, rather than a final project or one-time contribution. A collaborative environmental communication blog allows students to build from their experience with consumption of mediated messages about the environment (both the conscious critique during the course and more casual media viewing) to become constructors of environmental messages. In contrast to inert solo-authored blog assignments, a collaborative blog offers students the opportunity to consider and respond to a specific rhetorical exigence[1] collectively in a public sphere. It requires that students engage in significant tactical planning (delegating tasks, coordinating content, editing peers' work, etc.) that approximates the process and climate of working for an environmental campaign or grassroots organization. With guidance from the instructor, students can use current and cost-effective tools (such a Tumblr or Wordpress) to convey environmental messages to their intended public(s) in service of a specific mission.

The most recent edition of *Environmental communication and the public sphere* contains a new chapter addressing digital media in environmental activism (Cox and Pezzullo 2016). Collaborative blogs answer Cox and Pezzullo's (2016, 21) call for communication practitioners to "translate our private concerns into public matters and ... create circles of influence that affect how we and others view the environment and our relation to it." While it may seem difficult to achieve this in a single semester, collective action is more accessible through the use of digital communication technologies like blogs.

Incoming environmental communication students often adopt, adapt, and continue the blog constructed by previous semesters' students rather than establishing a discrete project. In essence, a collaborative blog provides a history of the environmental communication course and connects current students to past students in a way I have yet to achieve through other kinds of assignments. Some students are inspired to start their own blogging projects after the course, integrating lessons and tools from the semester into their lives beyond the classroom. For example, one student began a blog about integrating sustainability concerns into her Kosher lifestyle following the course.

Learning objectives

Students will:

- Choose and articulate a blog mission and target audience related to environmental communication.
- Construct the blog's design and multi-modal content to persuade their intended audience.
- Collaboratively plan, organize, and implement an editorial calendar.
- Evaluate peer-generated content and offer constructive feedback.
- Synthesize and address audience communication and public feedback to the blog.
- Analyze effectiveness of environmental messages in the public sphere and adapt when necessary.

Components of the project

Step 1: reading and critiquing other environmental blogs

If students are not avid readers of blogs, they may not know what constitutes success in writing for digital media. Consider introducing this assignment by requiring students to read a few different environmental blogs. This exploration will help them identify different missions and angles blogs have, as well as the diverse tones or structures science and environmental blogs can adopt. For example, the manner and charge of Andrew Revkin's (2015) "Dot Earth" blog is quite different from that of Mother Nature Network's (2015) "Treehugger," or Colin Beaven's (2015) now-archived "No impact man" blogging project. If this is an assignment you have used multiple times, consider having students read and critique the work of former classes.

Step 2: brainstorming a mission and target audience

Students' first collaborative task is coming to consensus about the scope and focus of their joint undertaking. Students will need to clarify what the purpose or mission of their blog is. Who are they trying to reach and why? Cox and Pezzullo (2016) distinguish between three objectives that environmental advocacy campaigns commonly have: they can alert, amplify, and/or engage. These objectives can serve as a useful framework for environmental communication students as they deliberate what their aims are for the project. Blogging for a particular rhetorical audience is another way to add a clear parameter to what might otherwise be an unwieldy assignment:

- Do they want to write for teenagers?
- For members of their campus community?
- For vegans looking for local resources?

During my environmental communication courses, students have often defined their target audience and mission locally, focusing on what Pittsburghers should know about climate disruption. One group decided to focus specifically on educating local teenagers about fossil fuel extraction and how individuals might take action to decrease fossil fuel use in the region. Grounding the blogging project in Lloyd Bitzer's (1968) concept of rhetorical situation is an excellent way to help students:

- focus on a particular exigence;
- call a rhetorical audience into being through their content; and
- think systematically about their constraints as environmental communicators.

If the class is large, it may be appropriate to divide into groups of five or six students for this assignment. If there are twenty or fewer in the course, consider having students all contribute to the same blog with the same general mission and audience.

Step 3: familiarizing students with the technological tools available

Many instructors assume our traditional college students are familiar and comfortable with social and new media technologies. However, this is not necessarily true. In order to increase the likelihood that students succeed in collaborative blogging, consider devoting some class time to exploring blogging tools (Wordpress, Blogger, Tumblr, and Typepad are all good free tools.) Most institutions have support specialists who can take this role if the instructor prefers.

Step 4: defining roles of group members and planning an editorial calendar

Once students know what is possible using blogging platforms, and what other environmental messages are being framed and addressed via blogs, it is time for them to determine who will fill which roles and how and when they will craft content. An inherently practical discipline, environmental communication should not shy away from giving students logistical problems to address. To help organize submissions and make sure all content is not posted at once there should be a set of student editors in charge of each sub-category of content posted to the environmental blog (e.g., News, Interviews, Local, Food, Science, Technology.) Students need to draft and adopt editorial guidelines for the blog, as well as establish a process for editing one another's work and scheduling posts to go live. To help them develop these workflow processes, consider showing some models from other collaborative blogging projects. One great model is Shout Out! JMU (2015), a student-run feminist blogging project that operates year round.

Such an undertaking works best as a semester-long blog, rather than a short-term project. While I recognize the moving pieces involved in a project like this might seem overwhelming, it is great practice for students in planning and implementing a collaborative environmental communication project. In my experience, students are often quick to judge the failed efforts of sustainability activists or community leaders early in the semester, but, once they experience for themselves how complex and delicate collaboration can be, they become capable of more nuanced and reflexive analysis of current events and environmental campaigns. The instructor should set a few parameters, such as the word count range for posts, how many posts each student is responsible for, requiring editor/peer feedback before a draft is published, clarifying how students can legally use images or videos in the blog, and the start and end dates for the project. To even out work distribution, students who take on editor roles, blog design roles, or marketing roles are expected to generate less individual content for the blog.

Challenges and issues will come up over the course of the semester, and students will have to reflect and adjust accordingly. For example, when during the first semester I adopted an environmental communication course blog, students were stumped when they received their first negative comments from anonymous readers. They deliberated and ultimately decided on a comment moderation policy. Consider giving a small amount of class time each week for such discussions and trouble-shooting.

Step 5: post content and develop a publicity strategy

As instructors in environmental communication courses, we often seek to impress on our students that local action can have global impact and vice versa. However, it is a harsh reality for students that posting something on the Internet is no guarantee people will read it. Once students begin to post content to the blog, consider asking the group how they might guide their target audience to the site itself. Students who have taken courses in public relations, social media, marketing, search engine optimization, or advertising may have some interdisciplinary knowledge to contribute to the discussion. All students can reflect on how they find new things to read and what makes them share content. Consider using embedded analytic tools offered by the blogging software to analyze which of the blog's posts are most read or shared, and why. It is important, particularly in a pragmatic course such as environmental communication, for students to evaluate how effective their reach is. Consider having students set a goal for the semester for their blog's reach:

- Do they want 100 followers to their RSS feed?
- Will they measure success by how many comments they receive?

Students can experiment with different strategies for increasing readership to determine what works best. Some strategies my students have used to good effect include:

- promotion through other social media tools such as Facebook or Twitter;
- use of keyword tags to increase search engine optimization;
- collaborations with established bloggers; and
- promotion through the university marketing or public relations office.

Evaluation

There is a great opportunity with such a student-driven project to encourage student input on how the product is to be graded. Consider asking students to collaborate on the crafting of a rubric or allowing them to determine how points for the assignment will be allocated into different categories you supply (grammar and mechanics, persuasive argumentation, organization, use of multimodal content, faithfulness to student-declared mission, etc.)

Some questions to consider asking students early in the process include:

- What portion of the grade should be individual and what portion should be shared by the group?
- How would you like your work on this collaborative project to be judged?
- How would you know you have succeeded with your environmental blog?
- How should peer feedback and editor roles be allocated and graded as separate from post-author roles?

In my course, I incorporate student feedback into the crafting of a rubric to be applied to individual blog posts for grading. Additionally, the entire group earns points based upon how well they work together, the quality of the feedback they give one another on their work, adherence to deadlines, and how successful they are in reaching their intended audience. I err on the side of generosity in grading the group portion and include extensive narrative feedback. Individual posts are graded within a week of publication, so students can incorporate feedback into subsequent posts.

Note

1 A *rhetorical exigence* is a specific problem, or communicative challenge, that demands rhetoric in response (Bitzer 1968).

References

Beaven, C. (2015) "No impact man," [Blog]. Available at: http://noimpactman.typepad.com/ (accessed March 1, 2016).

Bitzer, L. F. (1968) "The rhetorical situation," *Philosophy & Rhetoric*, 1: 1–14.

Cox, R. and Pezzullo, P. (2016) *Environmental communication and the public sphere*, Sage Publications, Thousand Oaks.

Gladwell, M. (October 4, 2010) "Small change: Why the revolution will not be tweeted," *The New Yorker*. Available at: www.newyorker.com/magazine/2010/10/04/small-change-malcolm-gladwell (accessed March 1, 2016).

Golding, P. (1998) "Worldwide wedge: Division and contradiction in the global information infrastructure," in Marris, P. and Thornham, S., eds., *Media studies: A reader*, New York University Press, New York, pp. 802–15.

Mother Nature Network Holding Company (2015) "Treehugger." Available at: www.treehugger.com (accessed March 1, 2016).

Revkin, A. (2015) "Dot earth." Available at: http://dotearth.blogs.nytimes.com/?_r=0 (accessed March 1, 2016).

Shout Out! JMU (2015) "Shout Out! JMU: Your source for feminist discourse." Available at: http://shoutoutjmu.com/ (accessed March 1, 2016).

Index

For Product Safety Concerns and Information please contact our EU
representative GPSR@taylorandfrancis.com
Taylor & Francis Verlag GmbH, Kaufingerstraße 24, 80331 München, Germany

www.ingramcontent.com/pod-product-compliance
Ingram Content Group UK Ltd.
Pitfield, Milton Keynes, MK11 3LW, UK
UKHW020016200926
475702UK00014B/1906

* 9 7 8 1 1 3 8 3 9 3 5 0 9 *